Approximation and Modeling with B-Splines

Approximation and Modeling with B-Splines

Klaus Höllig
Universität Stuttgart
Stuttgart, Germany

Jörg Hörner
Universität Stuttgart
Stuttgart, Germany

Society for Industrial and Applied Mathematics
Philadelphia

Copyright © 2013 by the Society for Industrial and Applied Mathematics

10 9 8 7 6 5 4 3 2 1

All rights reserved. Printed in the United States of America. No part of this book may be reproduced, stored, or transmitted in any manner without the written permission of the publisher. For information, write to the Society for Industrial and Applied Mathematics, 3600 Market Street, 6th Floor, Philadelphia, PA 19104-2688 USA.

Trademarked names may be used in this book without the inclusion of a trademark symbol. These names are used in an editorial context only; no infringement of trademark is intended.

MATLAB is a registered trademark of The MathWorks, Inc. For MATLAB product information, please contact The MathWorks, Inc., 3 Apple Hill Drive, Natick, MA 01760-2098 USA, 508-647-7000, Fax: 508-647-7001, *info@mathworks.com*, *www.mathworks.com*.

Library of Congress Cataloging-in-Publication Data

Höllig, Klaus, author.
  Approximation and modeling with B-splines / Klaus Höllig, Universität Stuttgart, Stuttgart, Germany; Jörg Hörner, Universität Stuttgart, Stuttgart, Germany.
    pages cm. – (Applied mathematics)
  Includes bibliographical references and index.
  ISBN 978-1-611972-94-8 (alk. paper)
1. Spline theory. 2. Approximation theory. 3. Numerical analysis. 4. Mathematical models.
5. Engineering–Mathematical models. 6. Computer science–Mathematics. 7. Algorithms. 8. Spline theory–Industrial applications. I. Hörner, Jörg, author. II. Title.
  QA224.H645 2013
  511'.4223–dc23
                                    2013029695

siam is a registered trademark.

Contents

Preface	vii
Introduction	ix
Notation and Symbols	xiii

1 Polynomials — 1
- 1.1 Monomial Form — 1
- 1.2 Taylor Approximation — 3
- 1.3 Interpolation — 5
- 1.4 Bernstein Polynomials — 9
- 1.5 Properties of Bernstein Polynomials — 11
- 1.6 Hermite Interpolant — 14
- 1.7 Approximation of Continuous Functions — 16

2 Bézier Curves — 19
- 2.1 Control Polygon — 19
- 2.2 Properties of Bézier Curves — 22
- 2.3 Algorithm of de Casteljau — 25
- 2.4 Differentiation — 27
- 2.5 Curvature — 29
- 2.6 Subdivision — 31
- 2.7 Geometric Hermite Interpolation — 33

3 Rational Bézier Curves — 37
- 3.1 Control Polygon and Weights — 37
- 3.2 Basic Properties — 39
- 3.3 Algorithms — 42
- 3.4 Conic Sections — 45

4 B-Splines — 51
- 4.1 Recurrence Relation — 51
- 4.2 Differentiation — 55
- 4.3 Representation of Polynomials — 59
- 4.4 Splines — 61
- 4.5 Evaluation and Differentiation — 68
- 4.6 Periodic Splines — 73

5 Approximation 77
5.1 Schoenberg's Scheme 77
5.2 Quasi-Interpolation 81
5.3 Accuracy of Quasi-Interpolation 85
5.4 Stability 88
5.5 Interpolation 90
5.6 Smoothing 97

6 Spline Curves 105
6.1 Control Polygon 105
6.2 Basic Properties 110
6.3 Refinement 115
6.4 Algorithms 125
6.5 Interpolation 130

7 Multivariate Splines 133
7.1 Polynomials 133
7.2 Polynomial Approximation 136
7.3 Splines 138
7.4 Algorithms 142
7.5 Approximation Methods 145
7.6 Hierarchical Bases 150

8 Surfaces and Solids 155
8.1 Bézier Surfaces 155
8.2 Spline Surfaces 160
8.3 Subdivision Surfaces 164
8.4 Blending 166
8.5 Solids 169

9 Finite Elements 173
9.1 Ritz–Galerkin Approximation 173
9.2 Weighted B-Splines 177
9.3 Isogeometric Elements 180
9.4 Implementation 183
9.5 Applications 187

Notes and Comments 193

Appendix 197

Bibliography 205

Index 213

Preface

B-splines are fundamental to approximation and data fitting, geometric modeling, automated manufacturing, computer graphics, and numerical simulations. Because of their computational efficiency, flexibility, and elegance, B-spline techniques are much superior to other, more elementary, piecewise polynomial representations. As a consequence, they have become the method of choice in numerous branches of applied mathematics, computer science, and engineering.

In this book we give an introduction to the basic B-spline theory, describing approximation methods and algorithms as well as modeling and design techniques. We think that only a solid knowledge in all these areas provides an optimal basis for interdisciplinary research and handling of complex novel applications. The new finite element schemes with B-splines provide a perfect example of a successful synthesis of methods from these different fields.

Topics discussed in our book include the Bézier form, computing with B-splines, approximation and interpolation, spline representations of curves, surfaces, and solids, hierarchical bases, and finite element simulation. We do not aim for completeness as more comprehensive and specialized texts do. Instead, we focus on key results and methods which are most widely used in practice. In this way, every important aspect of B-spline theory is described in a relatively short monograph, leading from elementary basic material to advanced topics which are subject of current research.

The material of the book can be almost covered in a one-semester mathematics or computer science graduate course. The combination of mathematics, programming, modeling, and graphics makes the subject fascinating to teach. There is a never-ending supply of interesting thesis topics, typically provided by new industrial applications. Further stimulating the enthusiasm for B-splines among students and assisting in teaching are some of our goals. Together with this book we plan to provide

- a collection of problems, partially with solutions;

- slides for lectures;

- programs and demos.

This supplementary material will be made available on the website for our book (http://www.siam.org/books/ot132).

The book is essentially self-contained. Some basic facts from linear algebra, analysis, as well as elementary differential geometry and functional analysis, which are required, are listed in an appendix. Hence, the material is easily accessible not only for mathematics and computer science students but also for beginning graduates in engineering.

We have very much enjoyed writing this book. The Numerical Analysis and Geometric Modeling group in Stuttgart provided a very creative atmosphere. In particular, we gratefully acknowledge the excellent cooperation with Ulrich Reif and Joachim Wipper on finite element projects. Special thanks also to Elisabeth Höllig and Irmgard Walter for helping to proofread various drafts of our manuscipt.

Stuttgart, December 2012

Klaus Höllig and Jörg Hörner

Introduction

B-splines play an important role in many areas of applied mathematics, computer science, and engineering. Typical applications arise in approximation of functions and data, automated design and manufacturing, computer graphics, medical imaging, and numerical simulation. There are also beautiful results in pure mathematics, in particular on n-width [88] and in connection with box-spline theory [22]. This diversity of areas and techniques involved makes B-splines a fascinating research topic which has attracted a growing number of scientists in universities and industry.

Polynomial and piecewise polynomial approximations have been used in a number of different contexts for a very long time. But perhaps it is fair to say that the systematic analysis of splines began with Schoenberg's paper on approximation of equidistant data in 1946 [105]. His work initiated very active research on approximation methods, as is documented, e.g., in the early books on splines by Ahlberg, Nilson, and Walsh [1], de Boor [19], and Schumaker [111]. The full potential of B-splines for numerical methods was realized by de Boor. His algorithms for computing with linear combinations of B-splines remain basic tools for any spline software; see [15, 18] and *A Practical Guide to Splines* [19]. On the theoretical side, de Boor's definition of a multivariate B-spline [17] was the starting point of a genuine multivariate spline theory with key contributions by Dahmen and Micchelli (cf., e.g., [36] for a small sample of intriguing results on box-splines), de Boor, DeVore, Riemenschneider, Schumaker, and many others.

Parallel to the mathematical research on splines, engineers developed piecewise polynomial representations for automated design and manufacturing [46]. At Boeing, Ferguson [53] used spline interpolation for curve and surface modeling. At General Motors, Coons [33] invented the well-known blending scheme for curve networks. At Renault and Citroën, a fundamental step towards modern modeling techniques was made by Bézier [7, 8, 9, 10] and de Casteljau [25, 26]. They independently discovered the favorable geometric and algorithmic properties of the Bernstein basis, introducing control polygons as an intuitive tool for designers.

The conference in Utah [3], organized by Barnhill and Riesenfeld in 1974, helped to bridge the gap between the mathematical and engineering community. A new research area was founded: Computer Aided Geometric Design (CAGD). The famous knot insertion and subdivision algorithms for control polygons developed by Boehm [11] and Cohen, Lyche, and Riesenfeld [30] illustrate the beauty of the interplay between modeling (engineering) and approximation (mathematics). Farin's book [49] gives an excellent description of the emerging geometric spline theory (cf. also the recent books by Cohen, Riesenfeld, and Elber [31] and Prautzsch, Boehm, and Paluszny [90] as well as the *Handbook of Computer Aided Geometric Design* [52] and [48]).

B-splines also began to play an important role in computer graphics. Chaikin's algorithm [29, 100] was the first example of a fast spline-based rendering scheme for curves. Very well known are the algorithms of Catmull/Clark [27] and Doo/Sabin [43] for refinement of surface meshes with arbitrary topology (cf. the book by Peters and Reif [86] for the fascinating mathematical theory behind such seemingly simple subdivision strategies). Step by step, B-splines entered almost all engineering areas. Most recently, B-splines were used as finite elements. Weighted approximations, proposed by Höllig, Reif, and Wipper [69], and isogeometric methods, introduced by Hughes, Cottrell, and Bazilevs [71], have proved to be quite successful.

Λ splines, β-splines, ν-splines, ω-splines, τ-splines, A-splines, ARMA splines, B-splines, Bernoulli splines, BM-splines, box-splines, cardinal splines, Catmull-Rom splines, D^m-splines, Dirichlet splines, discrete splines, E-splines, elliptic splines, exponential Euler splines, exponential box splines, fundamental splines, g-splines, Gibbs-Wilbraham splines, $H^{m,p}$-splines, harmonic splines, Helix splines, Hermite splines, Hermite-Birkhoff splines, histosplines, hyperbolic splines, Inf-convolution splines, K-splines, L-monosplines, L-splines, Lagrange splines, LB-splines, Legendre splines, Lg-splines, M-splines, metaharmonic splines, minimal-energy splines, monosplines, natural splines, NBV-splines, NURBS, ODR splines, perfect splines, polyharmonic splines, Powell-Sabin splines, pseudo splines, Q-splines, Schoenberg splines, simplex splines, smoothing splines, super splines, thin-plate splines, triangular splines, trigonometric splines, Tschebyscheff splines, TURBS, v-splines, variation diminishing splines, vertex splines, VP-splines, web-splines, Wilson-Fowler splines, X-splines, ...

Different Types of Splines and Spline Techniques

To date more than 8000 articles and books on splines have been published. The above listing gives the impression that there is a spline for almost any application, and there is! Our book covers only a very limited selection—it serves as a basic introduction to B-spline theory. In addition to essential basic material, we have given priority to the description of topics which are of primary importance in practice and/or which are subject of continuing research.

Outline of the Text

After some preliminary discussion of basic facts about polynomials in Chapter 1, Chapters 2 and 3 are devoted to the Bernstein–Bézier form. This important special case allows us to explain many key features of B-spline representations in a relatively simple setting. Univariate B-splines are defined in Chapter 4. We take an algorithmic approach, deriving properties from the fundamental recurrence relations. Chapter 5 discusses various approximation methods such as interpolation, quasi-interpolation, and smoothing. Then in Chapter 6, we describe modeling techniques for spline curves. Fundamental geometric algorithms are knot insertion and uniform subdivision with numerous applications. Chapters 7 and 8 are devoted to the multivariate theory of approximation and modeling based on the tensor product formalism. This simplest and most efficient representation does also possess sufficient local flexibility in combination with hierarchical refinement. The last chapter gives a brief introduction to finite element simulation with weighted and isogeometric B-spline elements (cf. the books [65, 34] for a comprehensive treatment of these techniques).

Notation and Conventions

Throughout the book, linear combinations of B-splines,

$$p = \sum_k c_k b_{k,\xi}^n,$$

play the fundamental role. The letter "b" is a canonical choice for basis functions, and "c" stands for coefficients or control points. For simplicity, we will often suppress dependencies on parameters if they are fixed or implicitly understood while discussing a

Introduction

particular topic. For example, the standard univariate B-spline is denoted by

$$b_k = b_{k,\xi}^n,$$

where n is the degree and $(\xi_k, \ldots, \xi_{k+n})$ the knot vector.

Frequently, we will not explicitly specify ranges of summation. In this case, a sum $\sum_k f_k$ is taken over all k for which f_k is defined. For example, the linear combination of B-splines shown above involves all B-splines which correspond to the knot sequence ξ.

For spline functions, "x" is the preferred variable. Correspondingly, knots are denoted by ξ_k. For curves, "t", representing time, is the canonical parameter and τ_k is used for the knots. For the representation of surfaces and solids we also use t as parameter.

We use the symbol $\lesssim$ for inequalities valid up to a constant factor, i.e.,

$$f \lesssim g \Leftrightarrow f \leq c\,g$$

with an absolute constant c. The symbols $\gtrsim$ and $\asymp$ are defined analogously. Constants, which depend on parameters p_k are denoted by

$$c(p_1, p_2, \ldots).$$

The constants are generic, i.e., they may change in a sequence of inequalities.

Main results are highlighted by shaded boxes. Rather than using a conventional numbering, we refer to these statements by their description, e.g., "see the *Weierstrass Approximation Theorem* in Section 1.7". With few exceptions, a rigorous derivation is given in the text immediately following a theorem or algorithm. Moreover, examples illustrate the applications.

Comments on the history and references to the most relevant literature are given in the chapter "Notes and Comments" at the end of the book. With few exceptions, our book does not present new results. Therefore, especially concerning basic material, often several sources for a particular result are available. We have cited what appear to be historically the first articles and/or seem to best fit our approach for a particular topic. For an extensive bibliography, we refer to the excellent database of de Boor and Schumaker (cf. [24]).

Supplementary Material

On the website for our book (http://www.siam.org/books/ot132), we will develop a B-spline database. It is intended to assist professors in their course preparation and to help students with homework assignments and to become aquainted with B-spline algorithms. The supplementary material will consist of three parts.

Problems. We will provide a collection of problems with hints as well as programming assignments. A solution manual will be available to instructors on request.

Lectures. Beamer presentations for each topic of the book facilitate course preparation and provide handouts for students.

Programs. We implemented key algorithms in MATLAB [84]. These programs are not intended as a small software package meeting professional standards. Instead, we kept the code as simple as possible to make the details easily accessible to students. The demos will serve as illustrations of principal B-spline techniques.

Notation and Symbols

$\lesssim, \gtrsim, \asymp$	inequalities and equivalences up to constants
$\|\ \|_\infty, \|\ \|_{\infty,D}$	maximum norm of vectors, matrices, and functions
$\|\ \|$	absolute value, Euclidean norm
$\langle \cdot, \cdot \rangle$	scalar product
$a(\cdot, \cdot)$	bilinear form
$[a_0, \dots, a_k]$	convex hull
$\alpha_{k,\xi}^n$	coefficients of recursion for B-spline derivatives
b_k^n	Bernstein polynomial
$b_{k,\xi}^n$	B-spline
b^n	standard uniform B-spline
$b_{k,h}^n$	uniform B-spline with grid width h
$\mathbb{B}_h$	finite element subspace
$c(\cdot)$	constant, depending on parameters
c_k	B-spline coefficients, control points
D_ξ^n	parameter interval/hyperrectangle
dist	distance function
Δ	forward difference, Laplace operator
$\Delta(x_0, \dots, x_n)f$	divided difference
$\gamma_{k,\xi}^n$	coefficients of B-spline recursion
$H_\Gamma^1(D)$	Sobolev space
$\varkappa$	curvature
L_2	square integrable functions
∇	backward difference
$\mathbb{P}^n(D)$	polynomials
$\psi_{k,\xi}^n$	Marsden coefficients
Q	energy functional
S_ξ^n	spline space
$S_\xi^n(D)$	restricted spline space
$S_{\eta,T}^n$	periodic spline space
$S_\Xi^n(D)$	hierarchical spline space
ξ, τ	knot sequences
ξ_k^n, τ_k^n	knot averages

Chapter 1
Polynomials

Polynomials are fundamental to modeling and numerical methods. They provide canonical local approximations to smooth functions and are used extensively in geometric design. This chapter contains basic results which are particularly important for B-spline techniques.

After describing the standard monomial form in Section 1.1, we discuss the Taylor approximation and polynomial interpolation in Sections 1.2 and 1.3. Then we introduce in Sections 1.4 and 1.5 the Bernstein basis. This representation has favorable geometric and algorithmic properties, which are essential for modeling and design. As a first application, we construct piecewise cubic Hermite interpolants in Section 1.6. Moreover, we prove in Section 1.7 the Weierstrass approximation theorem which states that any continuous function can be obtained as limit of polynomials.

1.1 ▪ Monomial Form

The standard representation of a polynomial p is the monomial form, which is particularly well suited to describe the behavior of p near the origin.

> **Monomial Form**
>
> A (real) polynomial p of degree n is a linear combination of the monomials $x \mapsto x^k$, $k = 0, \ldots, n$:
> $$p(x) = c_0 + c_1 x + \cdots + c_n x^n,$$
> with coefficients $c_k \in \mathbb{R}$ and $c_n \neq 0$. The coefficients correspond to the derivatives at $x = 0$:
> $$k! c_k = p^{(k)}(0), \quad k = 0, \ldots, n.$$
>
>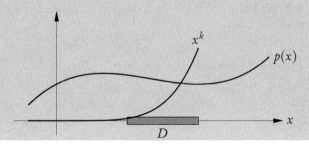

> The polynomials of degree $\leq n$ form a linear vector space of dimension $n+1$, denoted by $\mathbb{P}^n$. More precisely, we write $\mathbb{P}^n(D)$ if the variable x is restricted to a particular nondegenerate interval D.

The formula for the coefficients is easily checked. Since the kth derivative of a monomial $x \mapsto x^j$ vanishes for $j < k$ and equals

$$j(j-1)\cdots(j-k+1)x^{j-k}$$

for $j \geq k$, we have

$$p^{(k)}(x) = 0 + (k(k-1)\cdots 1)c_k + q(x),$$

where q is a linear combination of monomials with exponents ≥ 1. Setting $x=0$ yields the asserted expression for c_k.

Via differentiation we can easily show the linear independence of the monomials, proving that $\dim \mathbb{P}^n(D) = n+1$. Indeed, if

$$p(x) = \sum_{k=0}^{n} c_k x^k = 0, \quad x \in D,$$

then $p^{(n)}(x) = c_n n! = 0 \Rightarrow c_n = 0$. Forming successively the derivatives $p^{(n-1)}, p^{(n-2)}, \ldots$, we conclude that all coefficients are zero.

□ **Example:**

The graph of a linear polynomial p, given by

$$y = p(x) = c_0 + c_1 x,$$

is a straight line. As is illustrated in the figure, $(0, c_0)$ is the intersection with the y-axis and c_1 is the slope of p.

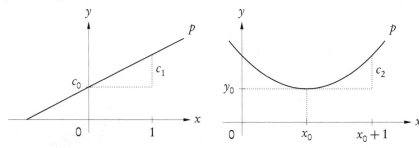

A quadratic polynomial parametrizes a parabola. By completing the square in the monomial form

$$p(x) = c_0 + c_1 x + c_2 x^2,$$

we obtain the standard representation

$$p(x) = c_2(x - x_0)^2 + y_0, \quad x_0 = -\frac{c_1}{2c_2}, \, y_0 = c_0 - \frac{c_1^2}{4c_2}.$$

□

When evaluating a polynomial in monomial form,

$$p(x) = c_0 + c_1 x + \cdots + c_n x^n,$$

taking powers can be avoided. To this end, we write p in nested form:

$$p(x) = c_0 + (c_1 + \cdots (c_{n-2} + (c_{n-1} + c_n x) x) x \cdots) x.$$

Starting with $p_n = c_n$, the value $p(x) = p_0$ is generated in n steps:

$$p_k = c_k + p_{k+1} x, \quad k = n-1, \ldots, 0.$$

This procedure is referred to as nested multiplication and requires $2n$ operations.

The recursion can also be used to compute the derivative of p. Differentiating the expression for p_k, we obtain $p'_n = 0$ and

$$p'_k = p'_{k+1} x + p_{k+1}, \quad k = n-1, \ldots, 0,$$

leading to $p'_0 = p'(x)$.

□ **Example:**
We illustrate nested multiplication for the example

$$p(x) = 4 - 3x + 2x^2 - x^3, \quad p'(x) = -3 + 4x - 3x^2.$$

Choosing, e.g., $x = 2$, the algorithm yields

k	3	2	1	0
c_k	-1	2	-3	4
p_k	-1	0	-3	-2
p'_k	0	-1	-2	-7

Hence, $p(2) = -2$ and $p'(2) = -7$. □

1.2 ▪ Taylor Approximation

The Taylor polynomial is the basic local approximation for smooth functions. It provides an error estimate which is used in many applications and numerical techniques.

Taylor Polynomial
The Taylor polynomial p_n of degree $\leq n$ of a function f at a point x_0 matches the derivatives up to order n:

$$p_n(x) = \sum_{k=0}^{n} \frac{f^{(k)}(x_0)}{k!} (x - x_0)^k.$$

The approximation error or remainder can be expressed in the form

$$f(x) - p_n(x) = \frac{f^{(n+1)}(\xi)}{(n+1)!} (x - x_0)^{n+1}$$

with ξ a point between x and x_0. As a consequence, polynomials of degree $\leq n$ approximate smooth functions on an interval $[x_0 - h, x_0 + h]$ with the order $O(h^{n+1})$.

In order to compute the derivatives of the Taylor polynomial p_n, we note that

$$\left(\frac{d}{dx}\right)^j (x-x_0)^k \bigg|_{x=x_0}$$

is nonzero only if $j=k$ and equals $j!$ in this case. Hence, $p_n^{(j)}(x_0) = f^{(j)}(x_0)$, showing that the derivatives match.

The formula for the remainder follows from the identity

$$f(x) - p_n(x) = \frac{1}{n!} \int_{x_0}^x f^{(n+1)}(t)(x-t)^n \, dt$$

by applying the mean value theorem

$$\int \varphi(t) \underbrace{\psi(t)}_{\geq 0} \, dt = \varphi(\xi) \int \psi(t) dt.$$

We set $\varphi = f^{(n+1)}$ and $\psi(t) = (x-t)^n$ and, in view of $\int_{x_0}^x (x-t)^n \, dt = (x-x_0)^{n+1}/(n+1)$, obtain the desired expression.

To derive the integral representation of the remainder $f - p_n$, we use induction on the degree. For $n = 0$, the identity follows from the fundamental theorem of calculus:

$$f(x) - f(x_0) = [f(t)]_{x_0}^x = \int_{x_0}^x f'(t) dt.$$

For the induction step from n to $n+1$, we integrate the right side of the identity for $f(x) - p_n(x)$ by parts and obtain

$$-\left[\frac{1}{(n+1)!} f^{(n+1)}(t)(x-t)^{n+1}\right]_{t=x_0}^{t=x} + \frac{1}{(n+1)!} \int_{x_0}^x f^{(n+2)}(t)(x-t)^{n+1} \, dt.$$

Bringing the first term,

$$\frac{f^{(n+1)}(x_0)}{(n+1)!} (x-x_0)^{n+1} = p_{n+1}(x) - p_n(x),$$

to the left side yields the desired form of the identity for degree $n+1$.

Finally, the error

$$\max_{|x-x_0| \leq h} |f(x) - p_n(x)|$$

can be estimated by ch^{n+1}, where $[(n+1)!c]$ is a bound of the absolute value of the $(n+1)$th derivative of f.

1.3. Interpolation

☐ **Example:**
The figure shows the first few Taylor polynomials at $x_0 = 0$ for the sine function:

$$p_1(x) = p_2(x) = x, \quad p_3(x) = p_4(x) = x - \frac{1}{6}x^3, \ldots$$

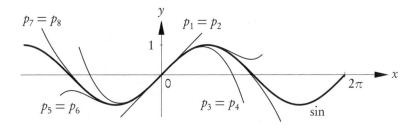

In accordance with the estimate for the remainder, the accuracy deteriorates as $|x|$ becomes large. Nevertheless, the approximations are globally convergent:

$$|\sin(x) - p_n(x)| \leq \frac{1}{(n+1)!}|x|^{n+1} \to 0 \quad (n \to \infty)$$

since the derivatives of the sine function are sines and cosines which are bounded by 1. ☐

In many applications, the function to be approximated is not given explicitly. Still, it can be possible to determine the Taylor polynomials. As a concrete example, we consider the initial value problem

$$y'(x) = x + y(x)^2, \quad y(1) = 0.$$

Differentiating, we obtain

$$y'' = 1 + 2yy', \quad y''' = 2y'y' + 2yy''.$$

Substituting the initial value at $x = 1$, we compute successively

$$y'(1) = 1, \quad y''(1) = 1, \quad y'''(1) = 2.$$

This yields

$$y(x) \approx (x-1) + \frac{1}{2}(x-1)^2 + \frac{2}{6}(x-1)^3$$

as an approximation near $x = 1$.

1.3 ▪ Interpolation

Polynomials can be used to predict the value of a function f at a point x from values at nearby points. The data are interpolated by a polynomial p which serves as an approximation to f. The classical explicit formula for this interpolant is attributed to Lagrange although first published by Waring in 1779.

Lagrange Form of Interpolating Polynomials

Any values f_k at $n+1$ distinct points x_k can be interpolated uniquely by a polynomial p of degree $\leq n$:

$$p(x_k) = f_k, \quad k = 0, \ldots, n.$$

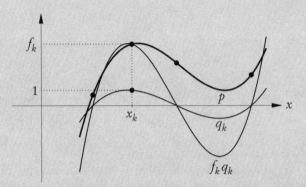

The Lagrange form of the interpolant is

$$p(x) = \sum_{k=0}^{n} f_k\, q_k(x), \quad q_k(x) = \prod_{\ell \neq k} \frac{x - x_\ell}{x_k - x_\ell},$$

with Lagrange polynomials q_k, which are equal to 1 at x_k and vanish at all other interpolation points x_ℓ, $\ell \neq k$.

By definition of the Lagrange polynomials q_k,

$$q_k(x_\ell) = \delta_{k,\ell}, \quad 0 \leq k, \ell \leq n,$$

with δ the Kronecker symbol. Hence,

$$p(x_\ell) = \sum_k f_k\, \delta_{k,\ell} = f_\ell,$$

i.e., p interpolates f_ℓ at x_ℓ.

If $\tilde{p}$ is another polynomial which interpolates the data (x_k, f_k), then $p - \tilde{p}$ has at least $n+1$ zeros. Since the difference is a polynomial of degree $\leq n$, it must vanish identically, showing the uniqueness of the interpolant.

□ **Example:**

Low degree polynomial interpolation is frequently used to generate intermediate function values. A typical example is the approximation

$$f_{k+1/2} \approx (-f_{k-1} + 9f_k + 9f_{k+1} - f_{k+2})/16,$$

which estimates $f(kh + h/2)$ from neighboring function values $f(\ell h)$ at the equally spaced points $x_\ell = \ell h$. The figure illustrates the simultaneous repeated application of this procedure for periodic data $f_k = \cos(\pi k) = (-1)^k$. After very few steps, the polygonal approximation obtained by connecting the generated points becomes visually smooth and approximates the cosine. Hence, the procedure yields a fast algorithm for plotting graphs of smooth functions based on uniformly sampled data.

1.3. Interpolation

The 4-point scheme is based on cubic interpolation. This means that the weights
$$-\frac{1}{16}, \frac{9}{16}, \frac{9}{16}, -\frac{1}{16}$$
are the values at $(k+1/2)h$ of the Lagrange polynomials corresponding to the four points $x_{k-1}, x_k, x_{k+1}, x_{k+2}$. For example,
$$-\frac{1}{16} = \left(\frac{x-kh}{(k-1)h-kh} \frac{x-(k+1)h}{(k-1)h-(k+1)h} \frac{x-(k+2)h}{(k-1)h-(k+2)h}\right)\bigg|_{x=kh+h/2}.$$

Because of symmetry, and since the weights must sum to 1 in order to reproduce constants ($f_{k-1} = \cdots = f_{k+2} = 1 \implies f_{k+1/2} = 1$), no further computations are necessary.

A theoretical analysis of the seemingly simple 4-point scheme is quite subtle. While the limit curve appears to be smooth, this is not the case. The second derivative ceases to exist, and approximations exhibit a fractal character. More precisely, as was shown by Dubuc, the first derivative of the limit function is Hölder continuous with exponent $1-\varepsilon$ for any $\varepsilon > 0$. □

When constructing polynomial interpolants, it is often necessary to add data in order to gain more accuracy. Moreover, successively increasing the degree also has algorithmic advantages. The interpolants can be generated with a simple triangular recursion.

Aitken–Neville Scheme

If p_k^{m-1} interpolates f at distinct points $x_k, \ldots, x_{k+m-1}$, then
$$p_k^m = (1-w_k^m)p_k^{m-1} + w_k^m p_{k+1}^{m-1}, \quad w_k^m(x) = \frac{x-x_k}{x_{k+m}-x_k},$$
interpolates at $x_k, \ldots, x_{k+m}$.

Starting with $p_k^0 = f(x_k)$, we obtain interpolating polynomials of successively higher degree with a triangular scheme:

$$\begin{array}{ccccccc}
p_0^0 & \to & p_0^1 & \to & p_0^2 & \cdots \to & p_0^n \\
p_1^0 & \nearrow & p_1^2 & \nearrow & & & \\
p_2^0 & \nearrow & & & & & \\
\vdots & & & & & & \\
p_n^0 & & & & & &
\end{array}$$

where the arrows $\to$ and $\nearrow$ pointing to p_k^m indicate multiplication with $(1-w_k^m)$ and w_k^m, respectively. The final polynomial p_0^n has degree $\leq n$ and interpolates at $x_0, \ldots, x_n$.

To derive the Aitken–Neville recursion, we check the interpolated values for each point x_ℓ in turn. For $x = x_k$, $w_k^m(x) = 0$, and

$$p_k^m(x_k) = 1 \cdot p_k^{m-1}(x_k) + 0 = f(x_k)$$

since p_k^{m-1} interpolates at x_k. Similarly, we conclude that $p_k^m(x_{k+m}) = f(x_{k+m})$. At the points x_ℓ, $\ell = k+1, \ldots, k+m-1$, p_k^{m-1} as well as p_{k+1}^{m-1} interpolate:

$$p_k^{m-1}(x_\ell) = f(x_\ell) = p_{k+1}^{m-1}(x_\ell).$$

Therefore,

$$p_k^m(x_\ell) = \left((1 - w_k^m(x_\ell)) + w_k^m(x_\ell)\right) f(x_\ell) = f(x_\ell),$$

as desired.

□ **Example:**
We illustrate the Aitken–Neville scheme by estimating $f(3)$ from the data

x_k	1	2	4
f_k	1	5	7

via quadratic interpolation. The triangular scheme yields

$$
\begin{array}{ccccc}
1 & \xrightarrow{2} & 9 & \xrightarrow{2/3} & 7 \\
 & & & \nearrow & \\
5 & \xrightarrow{1/2} & 6 & & \\
 & \nearrow & & & \\
7 & & & &
\end{array}
$$

where we have listed the weights $w_k^m(3)$ between the relevant arrows. For example,

$$7 = (1 - 2/3) \cdot 9 + (2/3) \cdot 6$$

is the approximation $p_0^2(3)$ of $f(3)$. □

Interpolation with polynomials of high degree is rarely used. In fact, even for very smooth functions, the interpolants need not converge as the degree tends to infinity. The following famous example is due to Runge.

□ **Example:**
The figure shows polynomial interpolants to the rational function $f(x) = 1/(1 + x^2)$ at equally spaced points in the interval $[-5, 5]$. While f is infinitely differentiable on $\mathbb{R}$, the interpolants diverge as the degree tends to infinity. Already for degree 10, the maximal error is about 2, as shown in the figure. The failure of polynomial interpolation is due to the singularities of f at $\pm i$ which are close to the real axis. Similarly as for Taylor series, poles in the complex plane limit the convergence interval for polynomial approximations.

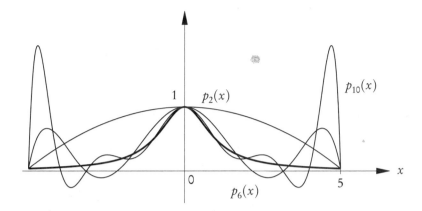

Runge's example indicates that polynomials are not a good choice for interpolating a large number of data. For such applications, splines yield substantially better results. □

1.4 ▪ Bernstein Polynomials

The monomial form is well suited to describe the local behavior of a polynomial near a specific point. However, this is not always of primary importance. Quite often it is preferable that the influence of the basis functions is more evenly distributed over the interval under consideration. Moreover, the basis should be symmetric with respect to the interval endpoints. A representation which meets these requirements was introduced by Bernstein.

> **Bernstein Polynomials**
> The Bernstein polynomials of degree n are defined by
> $$b_k^n(x) = \binom{n}{k}(1-x)^{n-k}x^k, \quad k=0,\ldots,n.$$
> They form a basis for the space $\mathbb{P}^n$ of polynomials of degree $\leq n$ which is symmetric with respect to the standard parameter interval $[0,1]$. In particular, for $j,k \in \{0,\ldots,n\}$,
> $$x^j = \sum_{k=j}^{n}\binom{k}{j}\bigg/\binom{n}{j}b_k^n(x), \quad b_k^n(x) = \sum_{j=0}^{n-k}(-1)^j\binom{n}{k}\binom{n-k}{j}x^{j+k},$$
> which describes the conversion between monomial and Bernstein form.
> In identities and recursions involving Bernstein polynomials it is often convenient to use indices $k \in \mathbb{Z}$ outside the range $\{0,\ldots,n\}$. For such k, $b_k^n(x)$ is set to zero, in agreement with a similar convention for binomial coefficients.

Both identities for the conversion between the two bases for $\mathbb{P}^n$ follow from the binomial theorem. Considering, e.g., the representation of the monomials in terms of

Bernstein polynomials, we have

$$x^j = x^j((1-x)+x)^{n-j}$$
$$= x^j \sum_{i=0}^{n-j} \binom{n-j}{i}(1-x)^{n-j-i}x^i.$$

Substituting $i = k - j$ gives

$$x^j = \sum_{k=j}^{n} \binom{n-j}{k-j}(1-x)^{n-k}x^k,$$

and, in view of

$$\binom{n-j}{k-j} = \binom{n}{k}\binom{k}{j} \Big/ \binom{n}{j}, \quad (1-x)^{n-k}x^k = b_n^k(x) \Big/ \binom{n}{k},$$

this yields the desired formula.

□ **Example:**
The figure shows the Bernstein polynomials b_k^n up to degree $n = 3$. Their explicit form is

$$\begin{array}{ll} b_k^0(x): & 1, \\ b_k^1(x): & 1-x, \quad x, \\ b_k^2(x): & (1-x)^2, \quad 2(1-x)x, \quad x^2, \\ b_k^3(x): & (1-x)^3, \quad 3(1-x)^2x, \quad 3(1-x)x^2, \quad x^3. \end{array}$$

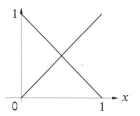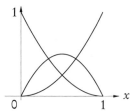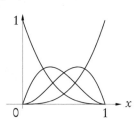

We see that the order of the zeros of b_k^n at the interval endpoints changes with k. For example, b_0^3 has a zero of order 3 at $x = 1$, b_1^3 has a zero at $x = 0$ and a zero of order 2 at $x = 1$, etc. □

With the aid of the identities for the basis functions, we can easily convert from monomial to Bernstein form and vice versa. We illustrate this for degree $n = 3$.

□ **Example:**
For cubic Bernstein polynomials

$$b_k^3: \quad (1-x)^3, 3(1-x)^2x, 3(1-x)x^2, x^3,$$

we have

$$\begin{pmatrix} 1 \\ x \\ x^2 \\ x^3 \end{pmatrix} = \begin{pmatrix} 1/1 & 1/1 & 1/1 & 1/1 \\ 0 & 1/3 & 2/3 & 3/3 \\ 0 & 0 & 1/3 & 3/3 \\ 0 & 0 & 0 & 1/1 \end{pmatrix} \begin{pmatrix} b_0^3(x) \\ b_1^3(x) \\ b_2^3(x) \\ b_3^3(x) \end{pmatrix}.$$

Writing the identity for the conversion from monomial to Bernstein basis in matrix form exhibits the pattern of binomial coefficients. In the numerator we recognize Pascal's triangle, and the denominators of the jth row equal $\binom{3}{j}$, $j = 0, \dots, 3$. Moreover, we can obtain the representation of an arbitrary polynomial by multiplying with the coefficient vector from the left. For example,

$$p(x) = 2 - 3x - x^3$$

has the Bernstein coefficients

$$\begin{pmatrix} 2 & -3 & 0 & -1 \end{pmatrix} \begin{pmatrix} 1 & 1 & 1 & 1 \\ 0 & 1/3 & 2/3 & 1 \\ 0 & 0 & 1/3 & 1 \\ 0 & 0 & 0 & 1 \end{pmatrix} = \begin{pmatrix} 2 & 1 & 0 & -2 \end{pmatrix},$$

i.e., $p(x) = 2b_0^3(x) + b_1^3(x) - 2b_3^3(x)$. $\qquad\square$

1.5 ▪ Properties of Bernstein Polynomials

In this section we describe basic properties of Bernstein polynomials and derive identities which will be crucial for subsequent applications.

Properties of Bernstein Polynomials

The Bernstein polynomials of degree n are nonnegative on the standard parameter interval $[0, 1]$ and sum to one:

$$\sum_{k=0}^{n} b_k^n(x) = 1.$$

Moreover, b_k^n has a unique maximum at

$$x = \frac{k}{n}$$

on $[0, 1]$.

At the interval endpoints 0 and 1, only the first and the last Bernstein polynomials are nonzero, respectively:

$$b_0^n(0) = 1, \quad b_1^n(0) = \cdots = b_n^n(0) = 0,$$
$$b_0^n(1) = \cdots = b_{n-1}^n(1) = 0, \quad b_n^n(1) = 1.$$

As a consequence, a polynomial in Bernstein form, $p = \sum_{k=0}^{n} c_k b_k^n$, is equal to c_0 at $x = 0$ and equal to c_n at $x = 1$. This property is referred to as endpoint interpolation.

By the binomial formula,

$$1 = (1 - x + x)^n = \sum_{k=0}^{n} \underbrace{\binom{n}{k}(1-x)^{n-k} x^k}_{b_k^n(x)}.$$

Hence, the Bernstein polynomials form a partition of unity.

The assertion about the maximum is obvious for $k = 0$ and $k = n$. For the other indices k, we compute
$$\frac{d}{dx} b_k^n(x) = \binom{n}{k}(1-x)^{n-k-1} x^{k-1}[-(n-k)x + k(1-x)].$$
Hence, the derivative of b_k^n vanishes for $x \in (0,1)$ iff the expression in brackets is zero:
$$0 = [\ldots] \Leftrightarrow x = \frac{k}{n}.$$
Since $b_k^n(0) = b_k^n(1) = 0$ for $k \in \{1, \ldots, n-1\}$ and b_k^n is positive on $(0,1)$, this parameter x must correspond to a unique maximum.

The assertions about the values at the interval endpoints are obvious from the definition of b_k^n.

☐ **Example:**

We can approximate a function f by using its values at the points where the Bernstein polynomials b_k^n are maximal as Bernstein coefficients:
$$f \approx p = \sum_{k=0}^{n} f(k/n) b_k^n.$$
As is illustrated in the figure, this so-called Bernstein approximation models the shape of the graph of f on $[0,1]$ quite accurately.

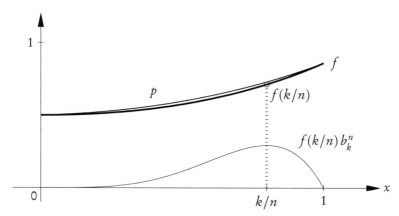

By the properties of Bernstein polynomials,
$$f(0) = p(0), \quad f(1) = p(1).$$
Moreover, positivity of f implies positivity of p. Finally, the approximation is exact for linear polynomials. Indeed, by the conversion formula for the monomial basis,
$$1 = \sum_k b_k^n, \quad x = \sum_k (k/n) b_k^n(x).$$
This means that $f = p$ for $f(x) = 1$ and $f(x) = x$ and, by linearity, also for $f(x) = c_0 + c_1 x$.

The simple approximation method was used by Bernstein in his proof of the Weierstrass approximation theorem, as will be described in Section 1.7. It has a canonical generalization to splines; cf. Section 5.1.   ☐

1.5. Properties of Bernstein Polynomials

Algorithms involving Bernstein polynomials are very elegant. This is a consequence of the following extremely simple formulas for basic operations.

Identities for Bernstein Polynomials

The Bernstein polynomials b_k^n, $k = 0, \ldots, n$, satisfy the following identities.

Symmetry:

$$b_k^n(1-x) = b_{n-k}^n(x).$$

Recursion:

$$b_k^n(x) = x\, b_{k-1}^{n-1}(x) + (1-x)\, b_k^{n-1}(x).$$

Differentiation:

$$\left(b_k^n\right)' = n\left(b_{k-1}^{n-1} - b_k^{n-1}\right).$$

Integration:

$$\int_0^1 b_k^n = \frac{1}{n+1}.$$

We note that $b_{-1}^{n-1} = b_n^{n-1} \overset{\centerdot}{=} 0$ in the second and third identities, according to the standard convention.

The symmetry of the Bernstein polynomials is easily seen from the definition. The recursion follows from the identity

$$\binom{n}{k} = \binom{n-1}{k-1} + \binom{n-1}{k}$$

for binomial coefficients. Indeed, adding

$$x\, b_{k-1}^{n-1}(x) = \binom{n-1}{k-1}(1-x)^{n-1-(k-1)}x^k$$

and

$$(1-x)\, b_k^{n-1}(x) = \binom{n-1}{k}(1-x)^{n-1-k+1}x^k$$

gives $\binom{n}{k}(1-x)^{n-k}x^k = b_k^n(x)$.

The differentiation formula is a direct consequence of the definition

$$\frac{d}{dx}\binom{n}{k}(1-x)^{n-k}x^k = \binom{n}{k}\left(k(1-x)^{n-k}x^{k-1} - (n-k)(1-x)^{n-k-1}x^k\right).$$

Noting that

$$k\binom{n}{k} = n\binom{n-1}{k-1}, \quad (n-k)\binom{n}{k} = n\binom{n-1}{k},$$

the assertion follows.

The integral of a Bernstein polynomial can be computed with the aid of the differentiation formula if we write it in the form

$$b_k^n = \frac{1}{n+1}\left(b_{k+1}^{n+1}\right)' + b_{k+1}^n.$$

Since, for $k = 0, \ldots, n-1$,

$$\int_0^1 \left(b_{k+1}^{n+1}\right)' = \left[b_{k+1}^{n+1}\right]_0^1 = 0,$$

it follows that

$$\int_0^1 b_0^n = \int_0^1 b_1^n = \cdots = \int_0^1 b_n^n.$$

Because the Bernstein polynomials sum to one, the common value of the integrals is $1/(n+1)$.

Alternatively, the integration formula can also be proved directly via integration by parts.

□ **Example:**

If many polynomials in Bernstein form have to be evaluated at the same points x_ℓ, the values

$$a_{\ell,k}^n = b_k^n(x_\ell)$$

should be precomputed. According to the identities for the Bernstein polynomials, the matrices A^n can be generated with the recursion

$$a_{\ell,k}^{n+1} = x_\ell\, a_{\ell,k-1}^n + (1-x_\ell)a_{\ell,k}^n,$$

where $a_{\ell,-1}^n = a_{\ell,n+1}^n = 0$. Then, the values

$$p(x_\ell) = \sum_{k=0}^n c_k\, b_k^n(x_\ell)$$

can be computed with the single matrix multiplication $A^n c$. □

1.6 ▪ Hermite Interpolant

The endpoint interpolation property facilitates the construction of interpolants. A particularly simple scheme is the cubic Hermite interpolant, which matches value and derivative at interval endpoints.

> **Hermite Interpolation**
>
> Values f_0, f_1 and derivatives d_0, d_1 at two points $x_0 < x_1$ can be interpolated by a cubic polynomial p. This Hermite interpolant can be expressed as linear combination of Bernstein polynomials transformed to the interval $[x_0, x_1]$:
>
> $$p(x) = f_0 b_0^3(y) + (f_0 + d_0 h/3)b_1^3(y) + (f_1 - d_1 h/3)b_2^3(y) + f_1 b_3^3(y)$$
>
> with $h = x_1 - x_0$ and $y = (x - x_0)/h$.

1.6. Hermite Interpolant

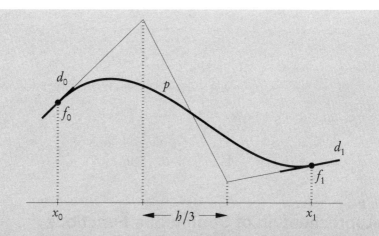

As illustrated in the figure, the Bernstein coefficients of the interpolant p form a polygon with equally spaced abscissae which touches p or, equivalently, the interpolated function at the interval endpoints.

If Hermite data are given at more than two points, the cubic interpolants form a so-called Hermite spline q. By construction, q is uniquely determined by its values and derivatives at the interpolation points and continuously differentiable there.

To prove the formula for the interpolant, we expand the Bernstein polynomials b_k^3 in powers of y:

$$p(x) = f_0\left(1 - 3y + O(y^2)\right) + (f_0 + d_0 h/3)\left(3y + O(y^2)\right) + O(y^2).$$

Since

$$x = x_0 \leftrightarrow y = 0, \quad \frac{d}{dx} = \frac{d}{dy} \underbrace{\frac{1}{h}}_{dy/dx},$$

it follows that

$$p(x_0) = f_0,$$
$$p'(x_0) = f_0(-3/h) + (f_0 + d_0 h/3)(3/h) = d_0,$$

as claimed. The argument for the right interval endpoint x_1 is analogous.

□ **Example:**

The figure shows the cross section of a vase modeled with the aid of piecewise Hermite interpolation.

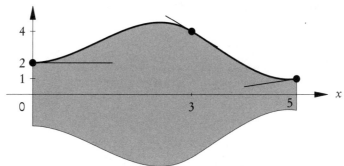

The top contour consists of two cubic segments. They are determined by the coordinates of the marked points and the specified slopes. For example, the data for the right segment are

$$(x_0, f_0) = (3, 4), (x_1, f_1) = (5, 1), \quad d_0 = -2, d_1 = 1/2.$$

By the formula for the Hermite interpolant, the corresponding polynomial is

$$p(x) = 4b_0^3(y) + (4 - 2 \cdot 2/3)b_1^3(y) + (1 - (1/2) \cdot 2/3)b_2^3(y) + b_3^3(y)$$

with $y = (x-3)/2$. Substituting the definition of b_k^3, a straightforward computation yields $p(x) = (3x^3 - 31x^2 + 89x - 37)/8$ for $x \in [3, 5]$. □

1.7 • Approximation of Continuous Functions

Polynomials not only provide very accurate approximations to smooth functions, but also guarantee convergence for any continuous function on a compact interval. This well-known result of Weierstrass can be proved using Bernstein polynomials with elementary arguments. In fact, the name for the basis functions b_k^n is based upon the usage of these polynomials in a proof by Bernstein.

> **Weierstrass Approximation Theorem**
> Any continuous function f can be approximated on a closed bounded interval $[a, b]$ by polynomials with arbitrary accuracy. More precisely, for any $\varepsilon > 0$, there exists a polynomial p with
> $$\max_{a \leq x \leq b} |f(x) - p(x)| < \varepsilon.$$

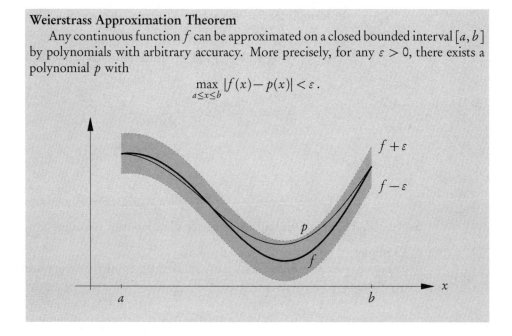

To derive Weierstrass's theorem, we may, without loss of generality, consider the interval $[a, b] = [0, 1]$. An elementary proof is based on the Bernstein approximation

$$f \approx p_n = \sum_{k=0}^{n} f(k/n) b_k^n.$$

Bernstein showed for this simple method that $p_n \to f$ as $n \to \infty$.

1.7. Approximation of Continuous Functions

To estimate the approximation error, we use that the Bernstein polynomials sum to one and write

$$f(x) - p_n(x) = \sum_{k=0}^{n} (f(x) - f(k/n)) \, b_k^n(x).$$

For a given $\varepsilon > 0$, we choose $\delta > 0$ so that

$$|x - y| < \delta \implies |f(x) - f(y)| < \varepsilon/2,$$

which is possible because of the uniform continuity of f on $[a, b]$.

We now split the summation indices $k \in \{0, \ldots, n\}$ into two groups:

$$I : |x - k/n| < \delta, \quad J : |x - k/n| \ge \delta.$$

This implies that

$$\left| \sum_{k \in I} (f(x) - f(k/n)) b_k^n(x) \right| < \frac{\varepsilon}{2} \sum_{k \in I} b_k^n(x) \le \frac{\varepsilon}{2}.$$

With c denoting the maximum of $|f|$ on $[0, 1]$, the other part of the sum in the expression of the error $f(x) - p_n(x)$ satisfies

$$\left| \sum_{k \in J} \cdots \right| \le \sum_{k=0}^{n} (2c) \left[\frac{(x - k/n)^2}{\delta^2} \right] b_k^n(x)$$

since the fraction in brackets is ≥ 1. As we will show below, the sum on the right side of the inequality can be computed explicitly:

$$\sum_{k} (x - k/n)^2 \, b_k^n(x) = \frac{x(1 - x)}{n}.$$

Hence, we can bound $|\sum_J \cdots|$ by

$$\frac{2c}{\delta^2} \frac{x(1 - x)}{n} \le \frac{2c}{\delta^2} \frac{1}{4n} < \frac{\varepsilon}{2},$$

provided that n is chosen sufficiently large. Summarizing, we obtain

$$|f(x) - p_n(x)| \le \left| \sum_{I} \cdots \right| + \left| \sum_{J} \cdots \right| < \frac{\varepsilon}{2} + \frac{\varepsilon}{2} = \varepsilon,$$

as claimed.

It remains to derive the formula for $x(1 - x)/n$. We use the Bernstein representation of the monomials:

$$x = \sum_{k} \frac{k}{n} b_k^n(x), \quad x^2 = \sum_{k} \frac{k(k - 1)}{n(n - 1)} b_k^n(x).$$

Forming a weighted sum, we also have

$$\frac{1}{n} x + \frac{n - 1}{n} x^2 = \sum_{k} (k/n)^2 \, b_k^n(x).$$

Using the above formulas, it follows that

$$\sum_k (x-k/n)^2 b_k^n(x) = x^2 \underbrace{\sum_k b_k^n(x)}_{=1} - 2x \underbrace{\sum_k (k/n) b_k^n(x)}_{=x} + \sum_k (k/n)^2 b_k^n(x)$$

$$= x^2 - 2x^2 + \left(\frac{1}{n}x + \frac{n-1}{n}x^2\right).$$

After simplification, this expression equals $x(1-x)/n$, as claimed.

☐ **Example:**

The figure illustrates the convergence of the Bernstein approximations p_n for the function $f(x) = |2x-1|\sqrt{x}$.

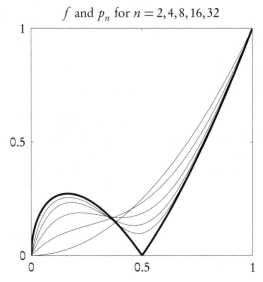

f and p_n for $n = 2, 4, 8, 16, 32$

In the neighborhood of $x = 0$ and $x = 1/2$, the convergence is rather slow. This is due to the singularities of the derivatives of f. For $x \to 0$, f' becomes infinite and, at $x = 1/2$, f' is discontinuous. As is to be expected, polynomials yield good approximations mostly in regions where f is smooth. ☐

Chapter 2

Bézier Curves

In the late 1960's Bézier and de Casteljau independently recognized the potential of Bernstein polynomials for engineering applications. The key observation is the geometric significance of the Bernstein coefficients. As is explained in this chapter, they serve as very intuitive control parameters for the computer aided design of curves. Moreover, the Bernstein basis has excellent numerical and algorithmic properties.

In Section 2.1, we introduce the parametrization of curves with Bernstein polynomials and in Section 2.2 derive basic properties of this representation referred to as Bézier form. Then, in Section 2.3, we describe de Casteljaus's evaluation scheme. This fundamental algorithm is also used to subdivide Bézier curves, as explained in Section 2.6. Section 2.4 is devoted to the differentiation of Bézier parametrizations. The simple formulas have many applications. In particular, we obtain an expression for the curvature and conditions for the smooth connection of Bézier curves in Section 2.5. Moreover, in Section 2.7, we describe a geometric Hermite interpolant, which yields extremely accurate approximations.

2.1 ▪ Control Polygon

If the coefficients in the Bernstein form have several components, we obtain a parametrization of a curve. This representation of polynomial curves, described in more detail below, has become a basic tool in computer aided geometric design and geometry processing.

> **Control Polygon**
> A Bézier curve p of degree $\leq n$ in $\mathbb{R}^d$ has a parametrization in terms of Bernstein polynomials:
> $$p(t) = \sum_{k=0}^{n} c_k b_k^n(t)$$
> with t in the standard parameter interval $[0,1]$.

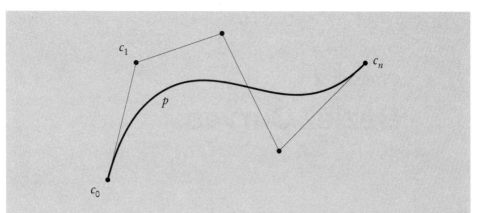

The coefficients $c_k = (c_{k,1}, \ldots, c_{k,d})$ can be combined into an $(n+1) \times d$ array C. They are called control points and form the control polygon c for p.

☐ **Example:**
A linear Bézier parametrization,

$$p(t) = c_0 \, b_0^1(t) + c_1 \, b_1^1(t) = c_0(1-t) + c_1 \, t,$$

represents the line segment $[c_0, c_1]$. As is illustrated in the left figure, the point $p(t)$ divides the segment in the ratio $t : (1-t)$.

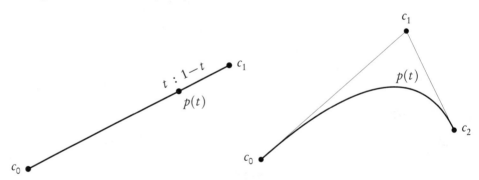

If the control points are not collinear, a quadratic Bézier parametrization,

$$p = \sum_{k=0}^{2} c_k \, b_k^2,$$

describes a piece of a parabola, as illustrated in the right figure. This is most easily seen by converting to monomial form:

$$p(t) = [c_0 - 2c_1 + c_2] t^2 + \{-2c_0 + 2c_1\} t + c_0.$$

For a quadratic curve, the coefficient of t^2 is nonzero and parallel to the axis of symmetry of the parabola. To obtain a normalized representation, we set

$$[\ldots] = \alpha u, \quad \{\ldots\} = \alpha' u + \beta v$$

with u and v orthonormal unit vectors. Changing the parametrization (not the curve) by the substitution

$$t = \frac{s}{\beta} - \frac{\alpha'}{2\alpha}$$

yields (after some elementary algebra)

$$p(t) = q(s) = \frac{\alpha}{\beta^2} s^2 u + s v + a, \quad a = c_0 - \frac{(\alpha')^2}{4\alpha} u - \frac{\alpha'\beta}{2\alpha} v,$$

with a the vertex of the parabola. In other words, if (x, y) are the coordinates with respect to the basis $\{u, v\}$ and origin at a, then

$$y = \gamma x^2, \quad \gamma = \alpha/\beta^2,$$

yielding the standard implicit form of a parabola. $\square$

Quadratic Bézier curves provide only limited design flexibility. However, already with parametrizations of degree 3, a number of qualitatively different shapes can be modeled, as is illustrated by the following example.

□ **Example:**
The figure shows planar cubic Bézier curves

$$p = \sum_{k=0}^{3} c_k b_k^3, \quad b_k^3(t) = \binom{3}{k}(1-t)^{3-k} t^k,$$

with control points $c_k = (c_{k,1}, c_{k,2})$ at the vertices of the unit square.

It is possible to let two or more control points coincide, which essentially results in the following possibilities.

The two right curves are straight line segments, though parametrized in a complicated way. For example,

$$p(t) = (0,0)(b_0^3(t) + b_1^3(t)) + (0,1)(b_2^3(t) + b_3^3(t)) = (0, 3t^2 - 2t^3)$$

describes the third curve. □

2.2 ▪ Properties of Bézier Curves

The geometric shapes of the control polygon and of the corresponding Bézier curve are very closely related. This is the consequence of a few simple properties summarized below.

Basic Properties
The shape of a Bézier curve parametrized by

$$p = \sum_{k=0}^{n} c_k b_k^n,$$

is qualitatively modeled by its control polygon c.

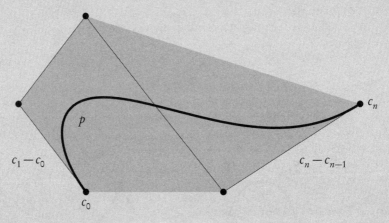

As illustrated by the figure,

- $p(t)$ lies in the convex hull of $c_0, \ldots, c_n$,
- $p(0) = c_0$, $p(1) = c_n$,
- $p'(0) = n(c_1 - c_0)$, $p'(1) = n(c_n - c_{n-1})$.

The last two properties referred to as endpoint interpolation imply that the control polygon is tangent to the Bézier curve, which is very useful for design purposes.

Since the Bernstein polynomials b_k^n are positive and sum to one, the convex hull property of Bézier curves is obvious.

2.2. Properties of Bézier Curves

To verify the interpolation property at the endpoints, we use that

$$b_k^n(0) = \delta_{k,0}, \quad b_k^n(1) = \delta_{k,n},$$
$$\left(b_k^n\right)' = n\left(b_{k-1}^{n-1} - b_k^{n-1}\right),$$

with δ the Kronecker symbol. Considering, e.g., the left endpoint, it follows that

$$p(0) = \sum_{k=0}^{n} c_k \delta_{k,0} = c_0$$

and

$$p'(0) = \sum_{k=0}^{n} c_k\, n(\delta_{k-1,0} - \delta_{k,0}) = n(c_1 - c_0),$$

as claimed.

☐ **Example:**
An important application of the convex hull property is the construction of bounding boxes. As is illustrated in the figure, the convex hull of the control points $c_0, \ldots, c_n \in \mathbb{R}^d$ is contained in the box

$$[c_1^-, c_1^+] \times \cdots \times [c_d^-, c_d^+],$$

where $c_\nu^- = \min_k c_{k,\nu}$ and $c_\nu^+ = \max_k c_{k,\nu}$.

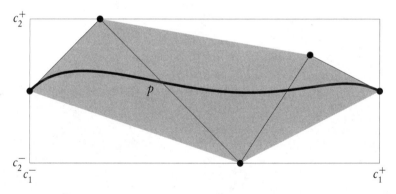

Bounding boxes are frequently used in numerical algorithms. A typical example is the location of curve intersections. Intersecting bounding boxes provides a quick test as to whether Bézier curves can have common points. ☐

Except for elementary cases, the description of geometric objects will involve more than one Bézier curve. Endpoint interpolation facilitates the construction of such piecewise Bézier representations. We just have to align the first and last edges of adjacent control polygons in order to achieve tangent continuity.

☐ **Example:**
The figure illustrates how to connect cubic Bézier curves. In this example, essentially only one control polygon is used, placed in different positions.

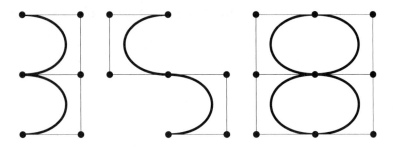

Degree 3 is preferred in practice because smooth connections can be constructed independently and four points per curve segment usually provide sufficient design flexibility. □

In view of endpoint interpolation, we can easily match Hermite data with Bézier parametrizations. However, there is a slight difference to the interpolation of functions, which we illustrate by the following example.

□ **Example:**
To achieve first order contact at two points p_0 and p_1 of a curve with an interpolating Bézier approximation $\sum c_k b_k^n$, it is sufficient to match the tangent directions. The derivatives of the parametrizations do not need to coincide, but their directions d_i have to:

$$c_0 = p_0, \qquad p_1 = c_n,$$
$$(c_1 - c_0) \parallel d_0, \qquad d_1 \parallel (c_n - c_{n-1}).$$

As illustrated in the figure, for a planar convex curve segment, $n = 2$ suffices rather than degree 3 as for the standard Hermite interpolant for functions. The middle control point c_1 of the quadratic approximation is obtained as the intersection of the two tangent lines.

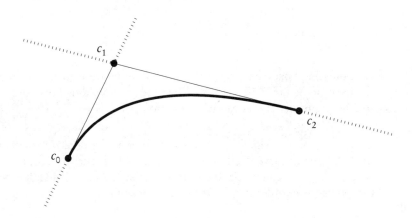

For smooth curves, the simple scheme is fairly accurate. For example, for a segment of a circle of length φ, the error is of order $O(\varphi^4)$. □

2.3 • Algorithm of de Casteljau

Points on a Bézier curve can be computed with a simple geometric procedure by repeatedly forming convex combinations of control points. This beautiful algorithm, due to de Casteljau, is the first example of many similar schemes operating on control polygons.

Algorithm of de Casteljau
A point
$$p(t) = \sum_{k=0}^{n} c_k \, b_k^n(t), \quad t \in [0, 1],$$
on a Bézier curve can be determined by successively subdividing the edges of the control polygon in the ratio $t : (1-t)$.

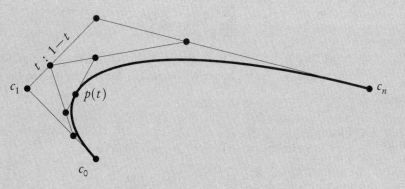

As is apparent from the figure, the computations can be arranged in a triangular scheme. The point $p(t)$ is obtained in n steps, each forming convex combinations of adjacent control points:
$$p_k^m = (1-t) p_k^{m-1} + t \, p_{k+1}^{m-1},$$
with $p_k^0 = c_k$ and $p_0^n = p(t)$.

To derive de Casteljau's scheme, we show that the intermediate points can be interpreted as points on Bézier curves corresponding to subsets of the control points:
$$p_k^m = \sum_{j=0}^{m} c_{k+j} \, b_j^m(t).$$

This yields a recursive version of de Casteljau's algorithm. The point $p(t)$ can be computed by combining points on two Bézier curves of degree $n-1$.

The formula for p_k^m is verified by induction. Assuming that

$$p_k^{m-1} = \sum_{j=0}^{m-1} c_{k+j}\, b_j^{m-1}(t), \quad k=0,\ldots,n-m+1,$$

we apply the subdivision step of the triangular scheme and obtain

$$p_k^m = (1-t)\sum_{j=0}^{m-1} c_{k+j}\, b_j^{m-1}(t) + t\sum_{j=0}^{m-1} c_{k+1+j}\, b_j^{m-1}(t).$$

Shifting the index in the second sum ($j \to j-1$), the right side equals

$$\sum_{j=0}^{m} c_{k+j}\left[(1-t)b_j^{m-1}(t) + t\, b_{j-1}^{m-1}(t)\right],$$

noting that $b_{-1}^{m-1} = b_m^{m-1} = 0$. By the recursion for Bernstein polynomials, the sum in brackets equals $b_j^m(t)$. Hence, we obtain the desired expression for p_k^m, completing the induction step.

☐ **Example:**

We consider the cubic Bézier curve parametrized by

$$p(t) = (-6,9)\, b_0^3(t) + (6,0)\, b_1^3(t) + (9,9)\, b_2^3(t) + (3,9)\, b_3^3(t), \quad t \in [0,1],$$

and compute $p(2/3)$ with de Casteljau's algorithm. The figure shows the three consecutive steps of the recursion, each dividing the edges of the control polygon in the ratio

$$t:(1-t) = 2/3 : 1/3.$$

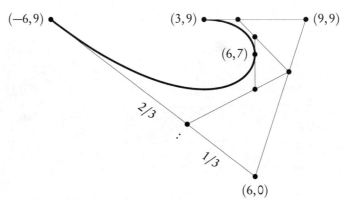

In the first step, the points

$$p_0^1 = (2,3),\ p_1^1 = (8,6),\ p_2^1 = (5,9)$$

are generated from the initial polygon. For example,

$$(2,3) = \frac{1}{3}(-6,9) + \frac{2}{3}(6,0).$$

In the final step, the point on the curve,

$$p(2/3) = p_0^3 = \frac{1}{3}(6, 5) + \frac{2}{3}(6, 8) = (6, 7),$$

is obtained. $\qquad\qquad\qquad\qquad\qquad\qquad\qquad\qquad\qquad\qquad\qquad\qquad\square$

2.4 ▪ Differentiation

The derivatives of a Bézier parametrization at the endpoints are proportional to differences of the first and last pair of control points. This is a special case of the following more general formulas which relate differentiation to a difference operation on the control point sequence.

> **Differentiation**
> The parametrization
>
> $$p = \sum_{k=0}^{n} c_k \, b_k^n$$
>
> of a Bézier curve is differentiated by forming differences of the control points:
>
> $$p' = n \sum_{k=0}^{n-1} (\Delta c_k) \, b_k^{n-1}$$
>
> with $\Delta c_k = c_{k+1} - c_k$. More generally, the mth derivative parametrizes a Bézier curve of degree $\leq n - m$ with control points
>
> $$\frac{n!}{(n-m)!} \Delta^m c_k, \quad k = 0, \dots, n-m.$$
>
> In particular,
>
> $$\binom{n}{m} \Delta^m c_0, \quad \binom{n}{m} \Delta^m c_{n-m}, \quad m = 0, \dots, n,$$
>
> are the Taylor coefficients at the endpoints.

The proof is a direct consequence of the formula for the derivative of a Bernstein polynomial,

$$\left(b_k^n \right)' = n \left(b_{k-1}^{n-1} - b_k^{n-1} \right).$$

It implies

$$p' = n \sum_{k=1}^{n} c_k \, b_{k-1}^{n-1} - n \sum_{k=0}^{n-1} c_k \, b_k^{n-1},$$

recalling that $b_{-1}^{n-1} = b_n^{n-1} = 0$. Shifting the index in the first sum ($k \to k+1$), the formula for the derivative of a Bézier parametrization follows. A repeated application of this argument yields the coefficients of higher derivatives.

In view of the endpoint interpolation property, the Taylor coefficients $p^{(m)}(t)/m!$ at $t = 0$ and $t = 1$ equal the first and last control point differences, respectively, multiplied by

$$\frac{1}{m!}\frac{n!}{(n-m)!} = \binom{n}{m}.$$

☐ **Example:**
As an example, we determine the Taylor coefficients of the cubic Bézier curve shown in the figure below.

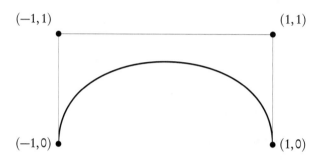

To this end, we first compute the control point differences conveniently arranged in a triangular array:

$$\begin{array}{lll}
\Delta c_0 = (0,1) & & \\
 & \Delta^2 c_0 = (2,-1) & \\
\Delta c_1 = (2,0) & & \Delta^3 c_0 = (-4,0). \\
 & \Delta^2 c_1 = (-2,-1) & \\
\Delta c_2 = (0,-1) & &
\end{array}$$

For example,

$$\Delta^2 c_1 = \Delta c_2 - \Delta c_1 = (0,-1) - (2,0),$$

etc.

Expanding at the left endpoint $t = 0$ gives

$$p(t) = \sum_{m=0}^{3} \binom{3}{m}(\Delta^m c_0) t^m$$
$$= (-1,0) + 3(0,1)t + 3(2,-1)t^2 + (-4,0)t^3$$
$$= (-1 + 6t^2 - 4t^3, 3t - 3t^2),$$

which of course agrees with the Taylor polynomial at $t = 1$:

$$p(t) = (1,0) + 3(0,-1)(t-1) + 3(-2,-1)(t-1)^2 + (-4,0)(t-1)^3. \qquad \square$$

Derivatives are needed in many numerical algorithms. As an illustration, we describe the computation of the distance to a Bézier curve.

☐ **Example:**
A closest point $p(t) = \sum_k c_k b_k^n(t)$ on a Bézier curve to a point q is either one of the endpoints ($t = 0$ or $t = 1$) or satisfies the orthogonality condition

$$\varphi(t) = \langle q - p(t), p'(t) \rangle = 0$$

2.5. Curvature

with $\langle \cdot, \cdot \rangle$ denoting the scalar product of two vectors. Hence, a numerical solution just requires determining the roots of the polynomial φ.

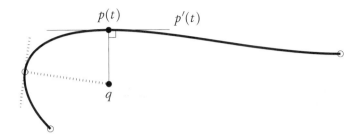

The polynomial φ has at most degree $2n-1$. It can be generated in four steps. First, we compute the control points $n\Delta c_k$ of p'. Then, we evaluate $p(t)$ and $p'(t)$ at $2n$ points, e.g., $t_\ell = \ell/(2n-1)$, $\ell = 0, \ldots, 2n-1$. Generating the values

$$\varphi(t_\ell) = \sum_{\nu=1}^{d} (q_\nu - p_\nu(t_\ell))\, p'_\nu(t_\ell),$$

we can finally determine the coefficients of φ by interpolation. □

2.5 ▪ Curvature

As for the tangent vectors, the control polygon also provides a simple geometric interpretation for the curvatures at the endpoints. The formulas involve the first and the last three control points, respectively. They can be used to characterize smooth connections of Bézier curves and to construct high precision interpolants.

Curvature

The curvatures $\varkappa$ at the endpoints of a Bézier curve, parametrized by

$$p = \sum_{k=0}^{n} c_k b_k^n,$$

have the following geometric interpretation. If $p'(0) \neq 0 \neq p'(1)$,

$$\varkappa(0) = \frac{2(n-1)}{n} \frac{\text{area}[c_0, c_1, c_2]}{|c_1 - c_0|^3}, \quad \varkappa(1) = \frac{2(n-1)}{n} \frac{\text{area}[c_n, c_{n-1}, c_{n-2}]}{|c_{n-1} - c_n|^3},$$

where $[a_0, a_1, a_2]$ denotes the triangle formed by the points a_k and $|v|$ is the length of a vector v.

To derive the formulas, we recall that the curvature of a curve with a regular parametrization p is given by

$$\varkappa = \text{area}(p', p'')/|p'|^3,$$

where area(f, g) denotes the area of the parallelogram spanned by the vectors f and g. Setting $u = c_1 - c_0$, $v = c_2 - c_1$, and $\alpha = n(n-1)$, the formulas for the first and second derivative yield

$$p'(0) = nu, \quad p''(0) = \alpha(v - u).$$

Substituting this into the definition of the curvature gives

$$\varkappa(0)|nu|^3 = \text{area}(nu, \alpha(v-u)) = n\alpha\,\text{area}(u, v)$$

by the rules for computing areas. Since area$(u, v) = 2\,\text{area}[c_0, c_1, c_2]$, we obtain the formula for $\varkappa(0)$.

The analogous identity for $\varkappa(1)$ is derived similarly.

☐ **Example:**

Bézier curves join smoothly if their control points are chosen appropriately. To describe the corresponding conditions, we consider two regular parametrizations $p^\pm$ with a common endpoint

$$c_n^- = p^-(1) = p^+(0) = c_0^+.$$

Continuous differentiability is equivalent to continuity of the unit tangent vector $p'/|p'|$. By the differentiation formula, this is the case if

$$c_1^+ - c_0^+ = \delta(c_n^- - c_{n-1}^-),$$

for some $\delta > 0$.

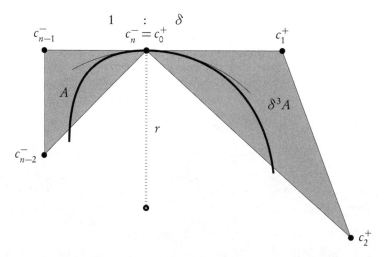

By elementary differential geometry, a curve is twice continuously differentiable at a point if it has second order contact with a circle or a line (which can be viewed as a degenerate circle). The corresponding conditions for a twice continuously differentiable connection of two Bézier curves are illustrated in the figure. The control points

$$c_{n-2}^-, c_{n-1}^-, c_n^- = c_0^+, c_1^+, c_2^+$$

relevant for the second order terms lie in the half plane containing the circle and bounded by the tangent line. Moreover, the curvatures match:

$$\varkappa^-(1) = 1/r = \varkappa^+(0),$$

where r is the radius of the osculating circle. By the formula for the curvature, the last condition is equivalent to

$$\delta^3 \, \text{area}[c_{n-2}^-, c_{n-1}^-, c_n^-] = \text{area}[c_0^+, c_1^+, c_2^+],$$

where δ is the ratio of the lengths of the tangent vectors. □

The geometric smoothness conditions were used by Boehm to construct planar curvature continuous spline curves. Two shape parameters for each edge of the control polygon provide additional flexibility. With these degrees of freedom smoothness can be further increased leading to four times continuously differentiable cubic curves. Unfortunately, this rather appealing mathematical discovery seems unpractical because of the highly nonlinear nature of the constraints.

2.6 ▪ Subdivision

Subdivision techniques are used in many numerical algorithms. For example, to locate intersections of Bézier curves, we can repeatedly split the curves at their midpoints and discard pairs which have nonintersecting bounding boxes.

The algorithm for subdivision uses de Casteljau's scheme. It not only computes a point on a Bézier curve but also generates the control points for the left and right curve parts.

Subdivision
A Bézier curve parametrized by

$$p(t) = \sum_{k=0}^{n} c_k \, b_k^n(t), \quad 0 \leq t \leq 1,$$

can be split into two parts corresponding to the subintervals $[0, s]$ and $[s, 1]$ with the aid of de Casteljau's algorithm.

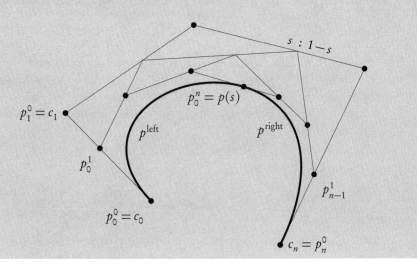

The first and last control points, p_0^m and p_{n-m}^m, generated with the mth de Casteljau step yield the control points of the left and right curve segments:

$$p^{\text{left}}(t) = p(st) = \sum_{m=0}^{n} p_0^m \, b_m^n(t),$$

$$p^{\text{right}}(t) = p(s+(1-s)t) = \sum_{m=0}^{n} p_m^{n-m} \, b_m^n(t).$$

Hence, the left and right control points correspond to the first and last diagonal of de Casteljau's scheme.

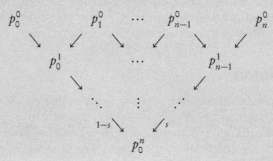

Considering, e.g., the left curve segment, we have to show that the scaled parametrization $(t \to st)$ equals the parametrization generated by the algorithm. In other words, we claim that

$$\sum_{k=0}^{n} c_k \, b_k^n(st) = \sum_{m=0}^{n} p_0^m(s) \, b_m^n(t),$$

where we have indicated the dependence of the points p_0^m on s. This identity is easily checked via induction starting from the trivial case $n = 0$.

For the induction step from n to $n+1$, we first observe that both sides of the above identity are polynomials of degree $\leq n$ in t and, hence, their derivatives at $t = 0$ match up to order n. By the differentiation formula and the chain rule, this implies that

$$s^\ell \Delta^\ell c_0 = \Delta^\ell d_0, \quad d_m = p_0^m(s),$$

for $\ell = 0, \ldots, n$. Therefore, the derivatives of the parametrizations $\varphi(t) = \sum_{k=0}^{n+1} c_k \, b_k^{n+1}(st)$ and $\psi(t) = \sum_{m=0}^{n+1} p_0^m(s) \, b_m^{n+1}(t)$ also match up to order n. Regardless of the degree, the formula for the ℓ-th derivative at $t = 0$ uses only the first $\ell + 1$ control points. Having a common Taylor polynomial up to degree n, the parametrizations differ by a term of order $n + 1$:

$$\varphi(t) - \psi(t) = a t^{n+1}.$$

In addition, both parametrizations agree at $t = 1$:

$$\varphi(1) = \sum_k c_k \, b_k^{n+1}(s) = p_0^{n+1}(s) = \psi(1)$$

by de Casteljau's algorithm. Hence, $a = 0$ and $\varphi = \psi$, which completes the induction step.

2.7. Geometric Hermite Interpolation

□ **Example:**
We consider the subdivision of quadratic and cubic Bézier curves at the midpoint $s = 1/2$ of the parameter interval.

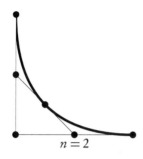
$n = 2$

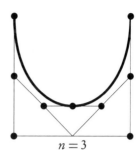
$n = 3$

In the quadratic case

$$c_1^{\text{left}} = \frac{1}{2}c_0 + \frac{1}{2}c_1,$$
$$c_2^{\text{left}} = \frac{1}{4}c_0 + \frac{1}{2}c_1 + \frac{1}{4}c_2,$$

where c_k^{left} denotes the control points of the left segment. Equivalently,

$$C^{\text{left}} = \begin{pmatrix} 1 & 0 & 0 \\ 1/2 & 1/2 & 0 \\ 1/4 & 2/4 & 1/4 \end{pmatrix} \underbrace{\begin{pmatrix} c_{0,1} & c_{0,2} & \cdots \\ c_{1,1} & c_{1,2} & \cdots \\ c_{2,1} & c_{2,2} & \cdots \end{pmatrix}}_{C}.$$

For the cubic case, the matrix form of the subdivision step is

$$C^{\text{left}} = \begin{pmatrix} 1 & 0 & 0 & 0 \\ 1/2 & 1/2 & 0 & 0 \\ 1/4 & 2/4 & 1/4 & 0 \\ 1/8 & 3/8 & 3/8 & 1/8 \end{pmatrix} C.$$

The general pattern is clear:

$$c_m^{\text{left}} = 2^{-m} \sum_{k=0}^{m} \binom{m}{k} c_k,$$

regardless of the degree of the Bézier curve. □

2.7 ▪ Geometric Hermite Interpolation

By Taylor's theorem, the error of local polynomial approximations for univariate functions is of order $O(h^{n+1})$, where n is the degree and h the length of the interval under consideration. Surprisingly, convergence rates are much better for curves. For example, it is conjectured that a smooth planar curve can, in general, be approximated with order $2n$ by Bézier curves of degree $\leq n$. The cubic geometric Hermite interpolant, proposed by de Boor, Höllig, and Sabin, was the first example of such high precision approximations.

Geometric Hermite Interpolation

The control points $c_0,\ldots,c_3$ of a planar cubic Bézier curve which interpolates the points f_j, the unit tangent directions d_j, and the signed curvatures $\varkappa_j$ ($j = 0,1$) at the endpoints $t = 0,1$ of the parameter interval satisfy

$$c_0 = f_0, c_3 = f_1, \quad c_1 = f_0 + \alpha_0 d_0/3, c_2 = f_1 - \alpha_1 d_1/3.$$

The lengths α_j of the tangent vectors are positive solutions of the nonlinear system

$$\varkappa_0 \alpha_0^2 = \det(d_0, 6(f_1 - f_0) - 2\alpha_1 d_1),$$
$$\varkappa_1 \alpha_1^2 = \det(d_1, 2\alpha_0 d_0 - 6(f_1 - f_0)).$$

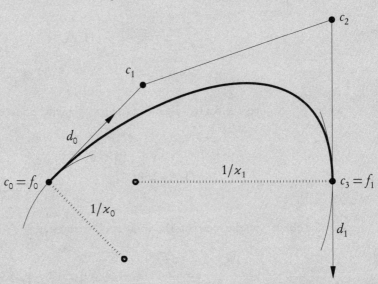

If the data correspond to a smooth curve with nonvanishing curvature, then for a sufficiently small distance $|f_1 - f_0|$, a solution of the nonlinear system exists and the error of the cubic Bézier approximation is of order $\mathcal{O}(|f_1 - f_0|^6)$.

The formulas for the coefficients c_k are an immediate consequence of the endpoint interpolation property of Bézier curves. For example, if p denotes the Bézier parametrization,

$$p(0) = c_0 = f_0, \quad \alpha_0 d_0 = p'(0) = 3(c_1 - c_0)$$

yield the formulas for c_0 and c_1, noting that $|p'(0)| = \alpha_0$.

The nonlinear system corresponds to the interpolation conditions for the signed curvatures. Considering, e.g., the left endpoint,

$$\varkappa_0 = \frac{\det(p'(0), p''(0))}{|p'(0)|^3} = \frac{\det(\alpha_0 d_0, 6((c_2 - c_1) - (c_1 - c_0)))}{\alpha_0^3}$$

by the definition of signed curvature and the formulas for the derivatives at endpoints. Substituting

$$c_2 - c_1 = (f_1 - f_0) - \alpha_1 d_1/3 - \alpha_0 d_0/3$$

and using that $c_2 - c_1 \| d_0$ and $\det(d_0, d_0) = 0$, we obtain the first equation of the nonlinear system.

The second equation is derived similarly.

The analysis of the error for smooth curves is rather subtle. The main difficulty is that the limit for $|f_1 - f_0| \to 0$ is a singular point of the system. Nevertheless, with the aid of the Newton polygon, a precise asymptotic analysis is possible. The following example illustrates some of the arguments in a trivial but instructive case.

□ **Example:**

We construct the geometric Hermite interpolant of a circular arc with radius 1 and opening angle 2ϑ. The symmetry of the data

$$\begin{aligned} f_0 &= (\cos\vartheta, -\sin\vartheta), & d_0 &= (\sin\vartheta, \cos\vartheta), & x_0 &= 1, \\ f_1 &= (\cos\vartheta, \sin\vartheta), & d_1 &= (-\sin\vartheta, \cos\vartheta), & x_1 &= 1, \end{aligned}$$

simplifies the computations, which can be carried out with the aid of a computer algebra system. The lengths α_j of the tangent vectors must be equal, and we have to consider only one equation of the nonlinear system:

$$\alpha^2 = \det(d_0, 6(f_1 - f_0) - 2\alpha d_1) = 12\sin^2\vartheta - 2\alpha\sin(2\vartheta).$$

There exists exactly one positive solution α of this quadratic equation which is a smooth function of ϑ:

$$\alpha = -\sin(2\vartheta) + \sqrt{\sin^2(2\vartheta) + 12\sin^2\vartheta} = 2\vartheta + \frac{1}{6}\vartheta^3 - \frac{7}{480}\vartheta^5 + O(\vartheta^7).$$

Substituting the expansion into the expressions for the control points of the Bézier parametrization (p_1, p_2), we can compute the deviation from the radius 1 of the circle:

$$1 - p_1(t)^2 - p_2(t)^2 = (t^3 - 3t^4 + 3t^5 - t^6)\vartheta^6 + O(\vartheta^7).$$

Hence, the error is of order $O(\vartheta^6)$ with $\vartheta \asymp |f_1 - f_0|$ illustrating the surprising gain in accuracy over standard cubic approximations. □

The remarkable phenomenon of improved approximation order persists for other examples. In general, it was conjectured by Höllig and Koch that Bézier curves of degree $\leq n$ approximate smooth curves in $\mathbb{R}^d$ with order

$$n + 1 + \left\lfloor \frac{n-1}{d-1} \right\rfloor.$$

For planar curves ($d = 2$), the gain over standard methods is $n-1$, as we remarked before.

Chapter 3
Rational Bézier Curves

A slight drawback of polynomial Bézier curves is that standard objects such as circles, ellipses, etc., cannot be represented exactly. With rational parametrizations, as introduced by Coons and Forrest to CAD/CAM applications, this is possible. However, the dependence on the parameters is no longer linear. This disadvantage is outweighed by additional design flexibility and improved compatibility with other geometric representations.

In Sections 3.1 and 3.2, we define the rational Bézier form and describe the basic properties of this more general curve format. We show in Section 3.3 how polynomial algorithms can be applied to manipulate rational parametrizations efficiently. Finally, Section 3.4 is devoted to the parametrization of conic sections.

3.1 ▪ Control Polygon and Weights

Like polynomial Bézier curves, rational Bézier curves are determined by a control polygon. In addition, each control point is given a positive weight controlling its influence on the shape of the curve.

Control Polygon with Weights

A rational Bézier curve r of degree $\leq n$ in $\mathbb{R}^d$ has a rational parametrization in terms of Bernstein polynomials:

$$r(t) = \frac{\sum_{k=0}^{n}(c_k w_k) b_k^n(t)}{\sum_{k=0}^{n} w_k b_k^n(t)}, \quad 0 \leq t \leq 1,$$

with positive weights w_k and control points $c_k = (c_{k,1}, \ldots, c_{k,d})$.

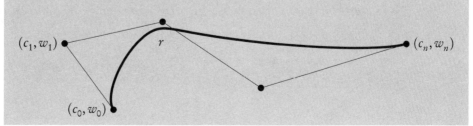

> As for polynomial Bézier curves, the control polygon c qualitatively models the shape of r. The weights give additional design flexibility by controlling the significance of the associated control points.

Scaling the weights, $w_k \to \lambda w_k$, does not change the parametrization of a rational Bézier curve. This extraneous degree of freedom can be eliminated by specifying merely the ratios $w_k : w_{k-1}$. As suggested by Farin, these ratios can be visualized as so-called weight points

$$d_k = \frac{w_{k-1}}{w_{k-1}+w_k} c_{k-1} + \frac{w_k}{w_{k-1}+w_k} c_k, \quad k=1,\ldots,n.$$

The position of d_k within the edge $[c_{k-1}, c_k]$ uniquely determines $w_k : w_{k-1} \in (0,\infty)$ and thus eliminates the inherent redundancy of the weights in an elegant fashion.

Moving adjacent weight points d_k, d_{k+1} towards the control point c_k increases the influence of this control point. The curve is pulled towards c_k.

Affine Invariance
  The parametrization

$$[0,1] \ni t \mapsto r(t) = \sum_{k=0}^{n} c_k \beta_k^n(t), \quad \beta_k^n = w_k b_k^n / \sum_{\ell=0}^{n} w_\ell b_\ell^n,$$

of a rational Bézier curve is affine invariant, i.e., if we apply an affine transformation

$$x \mapsto Ax + a$$

to r, we obtain the same result as with a transformation of the control points:

$$Ar + a = \sum_{k=0}^{n} (Ac_k + a)\beta_k^n.$$

3.2. Basic Properties

The affine invariance follows directly from

$$\sum_{k=0}^{n} \beta_k^n = 1.$$

This identity implies that

$$a = \sum_{k=0}^{n} a\,\beta_k^n$$

and, therefore, in view of the linearity of multiplication by A,

$$Ar + a = \sum_{k=0}^{n} \bigl(A(c_k \beta_k^n) + (a \beta_k^n)\bigr) = \sum_{k=0}^{n} (Ac_k + a)\beta_k^n,$$

as claimed.

Affine invariance is essential for design applications. For example, when transforming the coordinates of graphic windows, control polygons and the corresponding curves should change in the same fashion.

□ **Example:**

The figure illustrates the affine invariance for the elementary affine transformations: translation, shear, rotation, and scaling.

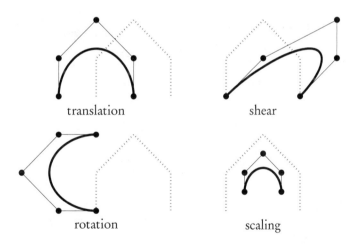

The examples confirm the intuitive geometric properties of Bézier curves. Any geometric operation on the control points yields the corresponding modification of the curve. □

3.2 ▪ Basic Properties

All elementary properties of polynomial Bézier curves have their counterparts for the more general rational representation. The weights merely cause some minor modifications in the basic formulas.

Properties of a Rational Bézier Curve
A rational Bézier curve parametrized by

$$r(t) = \frac{\sum_{k=0}^{n} (c_k w_k)\, b_k^n(t)}{\sum_{k=0}^{n} w_k\, b_k^n(t)}, \quad t \in [0,1],$$

has the following basic properties:

- $r(t)$ lies in the convex hull of the control points $c_0, \ldots, c_n$,

- $\lim\limits_{w_k \to \infty} r(t) = c_k$ for $t \in (0,1)$,

- $r(0) = c_0$, $r(1) = c_n$,

- $r'(0) = n \frac{w_1}{w_0}(c_1 - c_0)$, $r'(1) = n \frac{w_{n-1}}{w_n}(c_n - c_{n-1})$.

The weighted Bernstein polynomials

$$\beta_k^n = w_k b_k^n / q, \quad q = \sum_{\ell} w_\ell\, b_\ell^n,$$

are positive and sum to 1. Hence,

$$r(t) = \sum_k c_k\, \beta_k^n(t)$$

is a convex combination of the control points c_k, implying the convex hull property.
For $t \in (0,1)$, $b_k^n(t) > 0$ and

$$\lim_{w_k \to \infty} \beta_\ell^n(t) = \lim_{w_k \to \infty} \frac{(w_\ell / w_k)\, b_\ell^n(t)}{b_k^n(t) + \sum_{j \neq k}(w_j / w_k)\, b_j^n(t)} = \delta_{\ell,k},$$

proving the second property. Actually, a slightly stronger statement is true. As $w_k \to \infty$, $r(t)$ moves along a straight line. To see this, we compute

$$\frac{d}{dw_k} r(t) = c_k \frac{b_k^n(t)}{q(t)} - \sum_\ell c_\ell \frac{w_\ell b_\ell^n(t) b_k^n(t)}{q(t)^2} = (c_k - r(t)) \frac{b_k^n(t)}{q(t)},$$

confirming that the direction of change is parallel to $c_k - r(t)$.

The endpoint interpolation properties are verified by direct computation. Considering, e.g., the left endpoint $t = 0$ and using

$$b_k^n(0) = \delta_{k,0} = \beta_k^n(0), \quad \left(b_k^n\right)' = n(b_{k-1}^{n-1} - b_k^{n-1}),$$

we obtain

$$r(0) = \sum_k c_k\, \delta_{k,0} = c_0$$

3.2. Basic Properties

and

$$r'(0) = \sum_k c_k \frac{w_k n(b_{k-1}^{n-1}(0) - b_k^{n-1}(0))q(0) - w_k b_k^n(0)q'(0)}{q(0)^2}.$$

Since $q(0) = w_0$ and $q'(0) = n(w_1 - w_0)$, the sum simplifies to

$$c_1 \frac{w_1 n w_0}{w_0^2} + c_0 \frac{-w_0 n w_0 - w_0 n(w_1 - w_0)}{w_0^2},$$

which coincides with the desired expression for $r'(0)$.

□ **Example:**
The figure illustrates the influence of the weights for a planar rational cubic Bézier curve r with control points

$$C = \begin{pmatrix} 0 & 1 & 1 & 2 \\ 0 & 1 & -1 & 0 \end{pmatrix}^t.$$

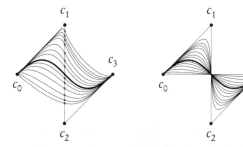
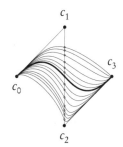

In the left example, starting from the polynomial Bézier curve ($w_k = 1$) in bold, the weight w_1 is increased ($w_1 \to \infty$) and decreased ($w_1 \to 0$). In the example on the right, w_2 is changed, while in the middle example $w_1 = w_2$ are modified simultaneously. □

Rational Bézier parametrizations are not unique. As described below, there are two degrees of freedom which can be eliminated by making a standard choice of the weights.

Parameter Transformation and Scaling
A rational Bézier curve parametrized by $\sum_k (c_k w_k) b_k^n / \sum_k w_k b_k^n$, is left invariant by scaling of the weights

$$w \to \lambda w$$

and by a linear rational parameter transformation of the form

$$t = \frac{s}{\varrho s + 1 - \varrho}, \quad \varrho < 1.$$

The two degrees of freedom can be used to change the first and last weight to 1,

$$w_k \to \tilde{w}_k = w_0^{k/n-1} w_n^{-k/n} w_k,$$

which is referred to as standard parametrization of a rational Bézier curve.

We consider the substitutions

$$w_k \to \lambda w_k, \quad t = s/\gamma(s),$$

with $\gamma(s) = \varrho s + 1 - \varrho$. For the Bernstein polynomials, this change of variables yields

$$b_k^n(t) = \binom{n}{k} \left(\frac{\varrho s + 1 - \varrho - s}{\gamma(s)} \right)^{n-k} \left(\frac{s}{\gamma(s)} \right)^k = \frac{(1-\varrho)^{n-k}}{\gamma(s)^n} b_k^n(s)$$

since $\varrho s + 1 - \varrho - s = (1-\varrho)(1-s)$. Hence, the parametrization of the rational Bézier curve becomes

$$\frac{\sum_k (c_k \lambda (1-\varrho)^{n-k} w_k) b_k^n(s)}{\sum_k (\lambda (1-\varrho)^{n-k} w_k) b_k^n(s)},$$

noting that the factors $\gamma(s)^n$ in numerator and denominator cancel. Defining the new weights

$$\tilde{w}_k = \lambda (1-\varrho)^{n-k} w_k,$$

the parametrization has Bézier form. Since $\varrho < 1$, the positivity of the weights is preserved.

Choosing

$$\lambda = 1/w_n, \quad (1-\varrho)^n = w_n/w_0,$$

we obtain the standard parametrization with $\tilde{w}_0 = \tilde{w}_n = 1$.

□ **Example:**

We consider the special case of a rational quadratic Bézier parametrization. According to the general formulas, the standard form of the parametrization is

$$r = \left(c_0 b_0^2 + (c_1 \tilde{w}_1) b_1^2 + c_2 b_2^2 \right)/q, \quad q = b_0^2 + \tilde{w}_1 b_1^2 + b_2^2,$$

with

$$\tilde{w}_1 = \lambda (1-\varrho) w_1 = w_1/\sqrt{w_0 w_2}$$

the only nontrivial weight. □

3.3 ▪ Algorithms

The implementation of algorithms for the rational Bézier form is facilitated by using homogeneous coordinates. As described below, we can apply the standard polynomial routines if we append the weights to the control points as an additional component.

Homogeneous Coordinates

The parametrization

$$t \mapsto r(t) = \frac{\sum_{k=0}^{n} (c_k w_k) b_k^n(t)}{\sum_{k=0}^{n} w_k b_k^n(t)}, \quad 0 \le t \le 1,$$

3.3. Algorithms

of a rational Bézier curve can be identified with a polynomial parametrization

$$\tilde{r} = (p \,|\, q) = \sum_k (c_k w_k \,|\, w_k) \, b_k^n$$

in homogeneous coordinates, i.e., $r = (p_1, \ldots, p_d)/q$. This interpretation is convenient for implementing algorithms such as evaluation, differentiation, and subdivision. The procedures for polynomial Bézier curves are applied to $\tilde{r}$, and the result in $\mathbb{R}^{d+1}$ is projected to $\mathbb{R}^d$ by dividing by the last coordinate.

☐ **Example:**

We evaluate the planar rational Bézier curve $r = p/q$ with

$$C = \begin{pmatrix} 0 & 0 \\ 4 & 0 \\ 0 & 2 \\ 4 & 3 \end{pmatrix}, \quad w = \begin{pmatrix} 2 \\ 2 \\ 3 \\ 2 \end{pmatrix}$$

at $t = 1/2$. Employing homogeneous coordinates, we apply de Casteljau's algorithm to the extended control points

$$\tilde{C} = \begin{pmatrix} w_0 c_0 & w_0 \\ w_1 c_1 & w_1 \\ w_2 c_2 & w_2 \\ w_3 c_3 & w_3 \end{pmatrix} = \begin{pmatrix} 0 & 0 & | & 2 \\ 8 & 0 & | & 2 \\ 0 & 6 & | & 3 \\ 8 & 6 & | & 2 \end{pmatrix}$$

which correspond to the space curve $\tilde{r} = (p_1, p_2 \,|\, q)$. The three steps of de Casteljau's triangular scheme yield

$$\tilde{C} \to \begin{pmatrix} 4 & 0 & | & 2 \\ 4 & 3 & | & 5/2 \\ 4 & 6 & | & 5/2 \end{pmatrix} \to \begin{pmatrix} 4 & 3/2 & | & 9/4 \\ 4 & 9/2 & | & 5/2 \end{pmatrix} \to \begin{pmatrix} 4 & 3 & | & 19/8 \end{pmatrix} = \tilde{r}(1/2).$$

Forming the quotient of the components $p(1/2) = (4,3)$ and $q(1/2) = 19/8$ of $\tilde{r}$ yields the midpoint $r(1/2) = (32/19, 24/19)$ of the planar curve.

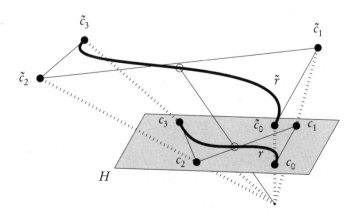

A geometric interpretation of the procedure is given in the figure. The control polygon c and the curve r can be identified with central projections of $\tilde{c}$ and $\tilde{r}$ onto the shaded plane

$$H = \{(r_1, r_2 \,|\, 1) \colon r_k \in \mathbb{R}\}.$$

Accordingly, $r(1/2)$ is the central projection of the point $\tilde{r}(1/2)$ generated by de Casteljau's algorithm. $\qquad\square$

Repeated differentiation of quotients results in complicated expressions. This can be avoided by a recursive computation of higher derivatives, as is explained below.

> **Differentiation**
>
> The parametrization
>
> $$r = \frac{\displaystyle\sum_{k=0}^{n} (w_k c_k)\, b_k^n}{\displaystyle\sum_{k=0}^{n} w_k\, b_k^n} = \frac{p}{q}$$
>
> of a rational Bézier curve can be differentiated with the aid of Leibniz's rule:
>
> $$\left(\frac{d}{dt}\right)^m \big(r(t) q(t)\big) = \sum_{\ell=0}^{m} \binom{m}{\ell} r^{(m-\ell)}(t) q^{(\ell)}(t) = p^{(m)}(t).$$
>
> This identity yields a recursion for $r^{(m)}$ in terms of the lower order derivatives:
>
> $$\begin{aligned}
> r' &= (p' - r q')/q, \\
> r'' &= (p'' - 2 r' q' - r q'')/q, \\
> r''' &= (p''' - 3 r'' q' - 3 r' q'' - r q''')/q, \\
> &\cdots.
> \end{aligned}$$
>
> For evaluating derivatives, we can therefore use the formulas and algorithms for standard Bézier curves. We simultaneously compute the derivatives of p and q and substitute the results into the recursion.

□ **Example:**

As an example, we compute the first and second derivative of $r(t)$ at $t = 0$, using the formulas

$$\tilde{r}'(0) = n(a_1 - a_0), \quad \tilde{r}''(0) = n(n-1)(a_2 - 2a_1 + a_0),$$

for the polynomial Bézier curve $(p \,|\, q)$ with control points $a_k = (c_k w_k \,|\, w_k)$.

For the first derivative

$$r' = (p' - r q')/q,$$

we obtain at $t = 0$

$$\begin{aligned}
r'(0) &= (n(c_1 w_1 - c_0 w_0) - c_0 n(w_1 - w_0))/w_0 \\
&= (n w_1 / w_0)(c_1 - c_0),
\end{aligned}$$

as mentioned in Section 3.2.

Similarly, evaluating the second derivative

$$r'' = (p'' - 2r'q' - rq'')/q$$

at $t = 0$ yields

$$r''(0) = (\alpha(c_2 w_2 - 2c_1 w_1 + c_0 w_0) - \beta(c_1 - c_0)(w_1 - w_0) - \alpha c_0(w_2 - 2w_1 + w_0))/w_0$$

with $\alpha = n(n-1)$ and $\beta = 2n^2 w_1/w_0$. After simplifications,

$$r''(0) = n(n-1)\frac{w_2}{w_0}(c_2 - c_1) - n\frac{2n w_1^2 - 2w_0 w_1 - (n-1)w_0 w_2}{w_0^2}(c_1 - c_0),$$

which is considerably more complicated than the expression

$$n(n-1)(c_2 - 2c_1 + c_0)$$

in the polynomial case, where $w_0 = w_1 = w_2 = 1$. $\qquad\square$

As a further illustration of the formulas for differentiation, we compute the curvature at the endpoints. By definition,

$$\varkappa = \operatorname{area}(r', r'')/|r'|^3.$$

Inserting the above formulas and noting that

$$\operatorname{area}(\gamma_1(c_1 - c_0), \gamma_2(c_2 - c_1) - \gamma_3(c_1 - c_0)) = |\gamma_1 \gamma_2| \operatorname{area}(c_1 - c_0, c_2 - c_1)$$

with $\gamma_1 = n w_1/w_0$ and $\gamma_2 = n(n-1)w_2/w_0$, we obtain after simplifications

$$\varkappa(0) = \frac{2(n-1)}{n}\frac{w_0 w_2}{w_1^2}\frac{\operatorname{area}[c_0, c_1, c_2]}{|c_1 - c_0|^3},$$

where $[c_0, c_1, c_2]$ denotes the triangle formed by the points c_k.

This formula and the analogous identity for the right endpoint differ from the expressions for the curvature of polynomial Bézier curves just by a factor involving the relevant weights.

3.4 ▪ Conic Sections

For the rational Bézier form, quadratic curves play a special role. In fact, the classical result, that any conic section admits a rational quadratic parametrization, has been the primary motivation for generalizing the polynomial representations.

Bézier Form of Conic Sections

Any rational quadratic Bézier curve parametrized by

$$r = \frac{(c_0 w_0) b_0^2 + (c_1 w_1) b_1^2 + (c_2 w_2) b_2^2}{w_0 b_0^2 + w_1 b_1^2 + w_2 b_2^2}$$

represents a segment of a conic section.

Conversely, any nondegenerate conic section can be represented by an extended parametrization $r(t)$, $t \in \mathbb{R} \cup \{\infty\}$.

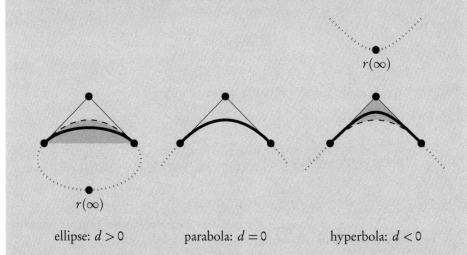

ellipse: $d > 0$ parabola: $d = 0$ hyperbola: $d < 0$

As is indicated in the figure, if the control points are not collinear, the type of the rational quadratic Bézier curve corresponds to the sign of $d = w_0 w_2 - w_1^2$.

If the control points are collinear, r represents a line segment (a point in the extreme case when all three control points coincide), i.e., a degenerate conic section. We exclude this trivial possibility in the sequel. Moreover, by affine invariance, we may assume that r is a planar curve. In effect, we consider a representation with respect to an orthogonal coordinate system of the plane spanned by the control points.

In homogeneous coordinates $(z_1, z_2 | z_3) \sim (x_1, x_2) = (z_1/z_3, z_2/z_3)$, a conic section Q is the zero-set of a quadratic form,

$$Q: \quad zAz^{\mathrm{t}} = 0,$$

with a symmetric 3×3 matrix A. To determine A from the parametrization r, we substitute

$$r = (p_1/q, p_2/q) \sim z = (p_1, p_2 | q)$$

into the implicit representation of the conic section. Since p_i and q have degree ≤ 2, we obtain an at most quartic polynomial in t:

$$\varphi(t) = \sum_{j,k=1}^{3} z_j(t) a_{j,k} z_k(t).$$

Its five coefficients depend linearly on the six entries

$$a_{1,1}, a_{2,2}, a_{3,3}, a_{1,2} = a_{2,1}, a_{1,3} = a_{3,1}, a_{2,3} = a_{3,2}$$

3.4. Conic Sections

of A. Hence, setting all coefficients to zero leads to a homogeneous 5×6-system for the matrix entries. Because of the dimensions, a nontrivial solution exists, which yields the implicit representation of r as a conic section.

To prove the converse, we note that any nondegenerate conic section can be obtained from the three normal forms

$$x_1^2 + x_2^2 = 1, \quad x_1^2 = x_2, \quad x_1 x_2 = 1$$

via an affine transformation. Extended quadratic rational parametrizations $r = (p_1, p_2)/q$ for the corresponding curves are easily constructed:

- circle:

$$r(t) = \frac{(1 - t^2, 2t)}{1 + t^2}, \quad (C \,|\, w) = \begin{pmatrix} 1 & 0 & 1 \\ 1 & 1 & 1 \\ 0 & 1 & 2 \end{pmatrix};$$

- parabola:

$$r(t) = \frac{(t, t^2)}{1}, \quad (C \,|\, w) = \begin{pmatrix} 0 & 0 & 1 \\ 1/2 & 0 & 1 \\ 1 & 1 & 1 \end{pmatrix};$$

- hyperbola:

$$r(t) = \frac{((2 - t)^2, (2 + t)^2)}{4 - t^2}, \quad (C \,|\, w) = \begin{pmatrix} 1 & 1 & 4 \\ 1/2 & 3/2 & 4 \\ 1/3 & 3 & 3 \end{pmatrix}.$$

It remains to verify the classification of the type of the conic section in terms of the weights. To this end, we observe that a transformation to standard form via

$$(w_0, w_1, w_2) \mapsto (\tilde{w}_0, \tilde{w}_1, \tilde{w}_2) = (1, w_1 / \sqrt{w_0 w_2}, 1)$$

leaves the sign of d invariant:

$$\tilde{d} = \underbrace{\tilde{w}_0 \tilde{w}_2}_{1} - \tilde{w}_1^2 = 1 - w_1^2 / (w_0 w_2) = d / (\underbrace{w_0 w_2}_{>0}).$$

Hence, we may assume that the denominator of r has the standard form

$$q(t) = b_0^2(t) + \tilde{w}_1 b_1^2(t) + b_2^2(t) = 1 + (\tilde{w}_1 - 1) b_1^2(t).$$

The type of r is now easily determined.

For $\tilde{d} > 0 \Leftrightarrow 0 < \tilde{w}_1 < 1$, q has no real zeros since $b_1^2(t) = 2t(1 - t) \le 1/2$. Therefore, $r(t)$ is bounded and approaches the same limit for $t \to \pm\infty$. The corresponding curve must be an ellipse.

For $\tilde{d} = 0 \Leftrightarrow \tilde{w}_1 = 1$, the denominator equals 1, i.e., $r(t) = p(t)$ parametrizes a parabola.

Finally, for $\tilde{d} < 0 \Leftrightarrow \tilde{w}_1 > 1$, q has two real zeros and therefore changes sign. Writing the numerator of r in the form

$$\begin{aligned} p &= c_0 b_0^2 + (c_1 \tilde{w}_1) b_1^2 + c_2 b_2^2 \\ &= (c_1 \tilde{w}_1) + (c_0 - c_1 \tilde{w}_1) \underbrace{b_0^2}_{\ge 0} + (c_2 - c_1 \tilde{w}_1) \underbrace{b_2^2}_{\ge 0}, \end{aligned}$$

we see that $p(t)$, $t \in \mathbb{R}$, lies in a cone with vertex $(c_1 \tilde{w}_1)$, spanned by the vectors $c_0 - c_1 \tilde{w}_1$ and $c_2 - c_1 \tilde{w}_1$. The sign change of the denominator q therefore results in two branches of the curve, which, consequently, must be a hyperbola.

□ **Example:**

We determine an implicit representation

$$(p_1, p_2 \,|\, q) \left(\begin{array}{cc|c} a_{1,1} & a_{1,2} & a_{1,3} \\ a_{1,2} & a_{2,2} & a_{2,3} \\ \hline a_{1,3} & a_{2,3} & a_{3,3} \end{array} \right) (p_1, p_2 \,|\, q)^{\mathrm{t}} = 0$$

of the quadratic rational Bézier curve, parametrized by $r = (p_1, p_2)/q$, with control points and weights

$$(C \,|\, w) = \left(\begin{array}{cc|c} 0 & 1 & 1 \\ 0 & 0 & 1/2 \\ 2 & 0 & 1 \end{array} \right).$$

Since $d = w_0 w_2 - w_1^2 = 1 - (1/2)^2 > 0$, we already know that r represents an ellipse. We now substitute

$$p_1(t) = 2t^2, \ p_2(t) = (1-t)^2, \ q(t) = (1-t)^2 + (1-t)t + t^2$$

into the implicit equation and obtain, with the aid of computer algebra,

$$\begin{array}{rl} (a_{2,2} + 2a_{2,3} + a_{3,3}) & \\ -(4a_{2,2} + 6a_{2,3} + 2a_{3,3}) & t \\ +(4a_{1,2} + 4a_{1,3} + 6a_{2,2} + 8a_{2,3} + 3a_{3,3}) & t^2 \\ -(8a_{1,2} + 4a_{1,3} + 4a_{2,2} + 6a_{2,3} + 2a_{3,3}) & t^3 \\ +(4a_{1,1} + 4a_{1,2} + 4a_{1,3} + a_{2,2} + 2a_{2,3} + a_{3,3}) & t^4 \end{array} = 0.$$

Setting the coefficients of t^k to zero and solving the resulting homogeneous linear system yields

$$a_{1,1} = 1, \ a_{1,2} = 1, \ a_{1,3} = -2, \ a_{2,2} = 4, \ a_{2,3} = -4, \ a_{3,3} = 4$$

as a possible solution. Hence, written out in detail, the implicit representation in homogeneous coordinates is

$$p_1^2 + 2p_1 p_2 - 4p_1 q + 4p_2^2 - 8p_2 q + 4q^2 = 0.$$

Dividing by q^2 yields the equation for the ellipse in cartesian coordinates,

$$x_1^2 + 2x_1 x_2 - 4x_1 + 4x_2^2 - 8x_2 = -4$$

with $x_k = p_k/q$. □

As a second example, we illustrate the construction of a Bézier parametrization from an implicit representation of a conic section.

3.4. Conic Sections

☐ **Example:**

To determine a rational quadratic parametrization $r = p/q$ of a conic section, we can select any two points bounding a convex curve segment, as control points c_0 and c_2. Because of endpoint interpolation, the middle control point c_1 then coincides with the intersection of the tangents. Finally, for a parametrization in standard form ($w_0 = 1 = w_2$), the nontrivial middle weight can be computed by testing one point of the parametrization.

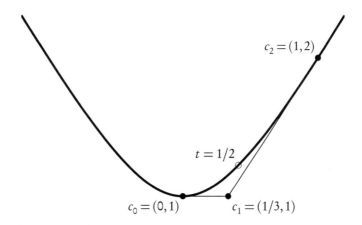

We illustrate the simple procedure for the hyperbola

$$Q: f(x) = 3x_1^2 - x_2^2 + 1 = 0.$$

As endpoints for the rational quadratic Bézier parametrization, we select $c_0 = (0, 1)$ and $c_2 = (1, 2)$. The equation of the tangent at the point $(1, 2)$ is

$$\text{grad } f(1,2)(x_1 - 1, x_2 - 2) = 6(x_1 - 1) - 4(x_2 - 2) = 0,$$

and we obtain $c_1 = (1/3, 1)$ as intersection with the horizontal tangent at $(0, 1)$. It remains to determine w_1. To this end, we evaluate

$$r = \frac{(p_1, p_2)}{q} = \frac{(0,1)b_0^2 + (1/3,1)w_1 b_1^2 + (1,2)b_2^2}{b_0^2 + w_1 b_1^2 + b_2^2}$$

at the parameter $t = 1/2$. Substituting

$$(x_1, x_2) = \frac{(p_1(1/2), p_2(1/2))}{q(1/2)} = \frac{(w_1/6 + 1/4, 1/4 + w_1/2 + 1/2)}{1/4 + w_1/2 + 1/4}$$

into the equation of the hyperbola Q and multiplying by $q(1/2)^2$, after simplification, yields the quadratic equation

$$\frac{1}{8} - \frac{1}{12} w_1^2 = 0$$

with the positive solution $w_1 = \sqrt{3/2}$. ☐

Chapter 4

B-Splines

Piecewise polynomial approximations are fundamental to many applications. However, using the monomial or Bernstein form, it is not straightforward to join the polynomial segments smoothly while keeping local flexibility. B-splines handle the smoothness constraints in a very elegant fashion and provide a basis with excellent numerical properties.

In Sections 4.1 and 4.2, we define B-splines and discuss the basic recurrence relations. Marsden's identity derived in Section 4.3 is the basis for the definition of the spline space in Section 4.4. Section 4.5 is devoted to evaluation and differentiation of spline functions. Finally, we consider periodic splines in Section 4.6.

4.1 ▪ Recurrence Relation

We begin by introducing some conventions for the partition of the parameter interval for piecewise polynomials. The concept of knot sequences with multiplicities will be convenient for specifying break points as well as smoothness constraints and is crucial for B-spline theory and algorithms.

Knot Sequence
A knot sequence

$$\xi : \cdots \leq \xi_0 \leq \xi_1 \leq \xi_2 \leq \cdots$$

is a finite or bi-infinite nondecreasing sequence of real numbers without accumulation points. It induces a partition of a subset $R \subseteq \mathbb{R}$ into knot intervals $[\xi_\ell, \xi_{\ell+1})$.

The multiplicity of a knot, $\#\xi_k$, is the maximal number of repetitions of ξ_k in the sequence ξ. In analogy with zeros of functions, we use the terms simple and double knots, etc.

Bi-infinite knot sequences are primarily of theoretical importance. In particular, they allow to handle index shifts in an elegant fashion. Moreover, periodic splines and global representations of polynomials require knot sequences which extend to $\pm\infty$.

Recurrence Relation

For a knot sequence ξ, the B-splines $b_{k,\xi}^n$ of degree n are defined by the recursion

$$b_{k,\xi}^n = \gamma_{k,\xi}^n b_{k,\xi}^{n-1} + (1-\gamma_{k+1,\xi}^n) b_{k+1,\xi}^{n-1}, \quad \gamma_{k,\xi}^n(x) = \frac{x-\xi_k}{\xi_{k+n}-\xi_k},$$

starting from the characteristic functions

$$x \mapsto b_{k,\xi}^0(x) = \begin{cases} 1 & \text{for } \xi_k \le x < \xi_{k+1}, \\ 0 & \text{otherwise}, \end{cases}$$

of the knot intervals $[\xi_k, \xi_{k+1})$ and discarding terms for which the denominator vanishes.

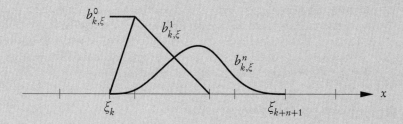

Each B-spline $b_{k,\xi}^n$ is uniquely determined by its knots $\xi_k, \ldots, \xi_{k+n+1}$ and vanishes outside of the interval $[\xi_k, \xi_{k+n+1})$. Moreover, on each of the nonempty knot intervals $[\xi_\ell, \xi_{\ell+1}), k \le \ell \le k+n$, it is a nonnegative polynomial of degree $\le n$.

If the degree and the knot sequence are fixed throughout the discussion of a particular topic, we write $b_k = b_{k,\xi}^n$ to avoid the excessive use of sub- and superscripts. Similarly, parameters are omitted in other B-spline related notations.

The basic properties of the B-spline $b_{k,\xi}^n$ are easily verified by induction.

The recursion implies that $b_{k,\xi}^n$ vanishes for $x < \xi_k$ and $x \ge \xi_{k+n+1}$ since $b_{k,\xi}^{n-1}$ and $b_{k+1,\xi}^{n-1}$ are zero for such arguments x. Moreover, $b_{k,\xi}^n \ge 0$ since both summands in the recursion are nonnegative. Finally, since $\gamma_{k,\xi}^n$ is a linear function, the recursion increases the polynomial degree at most by 1.

☐ **Example:**
For the knot sequence

$$(\xi_0, \xi_1, \xi_2, \xi_3) = (1, 2, 5, 7),$$

we determine the polynomial segments of all B-splines up to degree 2.

Starting with the characteristic functions $b_{k,\xi}^0$, $k = 0, 1, 2$, of the knot intervals

$$[1, 2), [2, 5), [5, 7),$$

4.1. Recurrence Relation

we obtain

$$b^1_{0,\xi}(x) = \frac{x-1}{2-1}b^0_{0,\xi}(x) + \frac{5-x}{5-2}b^0_{1,\xi}(x) = \begin{cases} x-1 & \text{for } 1 \leq x < 2, \\ \dfrac{5-x}{3} & \text{for } 2 \leq x < 5, \\ 0 & \text{otherwise}. \end{cases}$$

Hence, the linear B-spline $b^1_{0,\xi}$ is a so-called hat-function, which has the value 1 at the knot $\xi_1 = 2$ and vanishes at all other knots. Similarly, $b^1_{1,\xi}$ is a hat-function with value 1 at $\xi_2 = 5$.

For the quadratic B-spline $b^2_{0,\xi}$ the recursion is illustrated in the figure. This B-spline is nonzero on all three knot intervals, which we consider in turn. For $x \in [1,2)$, the linear B-spline $b^1_{1,\xi}$ is zero, and only $b^1_{0,\xi}$ contributes to the recursion:

$$b^2_{0,\xi}(x) = \frac{x-1}{5-1}(x-1) + 0 = \frac{1}{4}(x-1)^2.$$

For $x \in [2,5)$, both B-splines contribute:

$$b^2_{0,\xi}(x) = \frac{x-1}{5-1}\frac{5-x}{3} + \frac{7-x}{7-2}\frac{x-2}{3} = \frac{1}{20}(-3x^2 + 22x - 27).$$

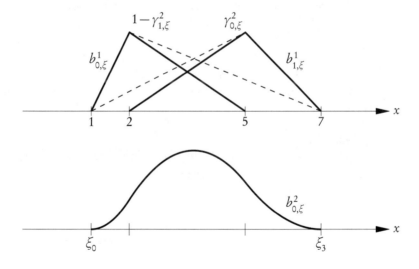

Similarly as for the left-most interval, we obtain

$$b^2_{0,\xi}(x) = \frac{1}{10}(7-x)^2$$

for the right-most interval, i.e., for $x \in [5,7)$. □

B-splines have similar properties as Bernstein polynomials. In fact, we recover the Bernstein basis b^n_k as a special case.

☐ **Example:**
The B-spline $b_{k,\xi}^n$ with the knots
$$0 = \xi_k = \cdots = \xi_n < \xi_{n+1} = \cdots = \xi_{k+n+1} = 1$$
coincides with a Bernstein polynomial:
$$b_{k,\xi}^n(x) = b_k^n(x) = \binom{n}{k}(1-x)^{n-k}x^k, \quad 0 \leq x < 1.$$
This justifies using a similar notation for the more general class of basis functions.

The relationship to Bernstein polynomials can be verified by induction. The case $n = 0$ is obvious, and the linear case is also easily to be checked. For example, the B-spline with a double knot at $x = 0$ and a simple knot at $x = 1$ coincides with the Bernstein polynomial $b_0^1(x) = (1-x)$ on $[0,1)$.

For the induction step from $n-1$ to n, we compute the B-spline $b_{k,\xi}^n$ with an $(n+1-k)$-fold knot at $x = 0$ and a $(k+1)$-fold knot at $x = 1$ with the recursion
$$b_{k,\xi}^n(x) = x\varphi(x) + (1-x)\psi(x),$$
where φ and ψ are B-splines of degree $n-1$ with knots $\xi_k, \ldots, \xi_{k+n}$ and $\xi_{k+1}, \ldots, \xi_{k+n+1}$, respectively. Since φ has an $(n+1-k)$-fold knot at $x = 0$ and a k-fold knot at $x = 1$, $\varphi = b_{k-1}^{n-1}$, by induction. Similarly, $\psi = b_k^{n-1}$, according to the knot multiplicities. Hence, the recursion for Bernstein polynomials shows that $b_{k,\xi}^n = b_k^n$ on $[0,1)$.

We emphasize that, because of the convention of continuity from the right, we cannot replace $[0,1)$ by $[0,1]$. For example, for $\xi = (0,0,0,1,1,1)$, $b_2^2(1) = x^2|_{x=1} = 1$ while $b_{2,\xi}^2(1) = 0$ in view of the triple knot. ☐

In principle, we can derive explicit formulas for the polynomial segments of a B-spline with the aid of the recurrence relation. However, simple expressions are only obtained in special cases. An example is the formula for the first and last segments.

☐ **Example:**
If $\xi_k < \xi_{k+1}$, then only the first term in the recursion for $b_{k,\xi}^n$ is nonzero on the interval $[\xi_k, \xi_{k+1})$. Hence, on this interval, the B-spline is a product of the factors $\gamma_{k,\xi}^m$:
$$b_{k,\xi}^n(x) = \frac{(x-\xi_k)^n}{(\xi_{k+1}-\xi_k)\cdots(\xi_{k+n}-\xi_k)}, \quad \xi_k \leq x < \xi_{k+1}.$$
An analogous formula is valid on the right-most interval $[\xi_{k+n}, \xi_{k+n+1})$ of the support of $b_{k,\xi}^n$. In particular, if $\xi_{k+1} = \cdots = \xi_{k+n}$, then, as is illustrated in the figure, the B-spline is made up of two monomials which are equal to 1 at the middle knot.

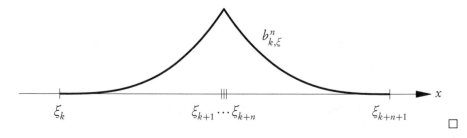

☐

4.2. Differentiation

We conclude this section with a simple but important remark about the dependence of a B-spline on the knots. As one expects, moving a knot results in a gradual change of a B-spline.

Continuous Dependence on the Knots
If x lies in the interior of one of the knot intervals of the B-spline $b_{k,\xi}^n$ and

$$\eta_\ell \to \xi_\ell, \quad \ell = k,\ldots,k+n+1,$$

then, as illustrated in the figure,

$$\lim_{\eta \to \xi} b_{k,\eta}^n(x) = b_{k,\xi}^n(x).$$

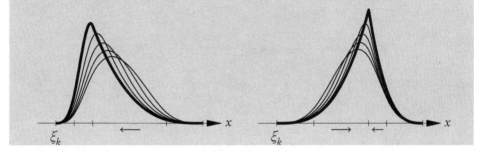

Aside from the obvious relevance for numerical stability, continuous dependence is often used in proofs. Arguments involving B-splines can be given for the usually somewhat less technical case of simple knots. By continuity, the results derived remain valid for multiple knots as well.

4.2 ▪ Differentiation

A second fundamental recursion for B-splines is the formula for the derivative. It allows us to determine the precise smoothness of $b_{k,\xi}^n$ at the knots.

Derivative of a B-Spline
The derivative of a B-spline of degree n with knots $\xi_k,\ldots,\xi_{k+n+1}$ is a weighted difference of two B-splines of degree $n-1$. On each knot interval $[\xi_\ell, \xi_{\ell+1})$,

$$\left(b_{k,\xi}^n\right)' = \alpha_{k,\xi}^n b_{k,\xi}^{n-1} - \alpha_{k+1,\xi}^n b_{k+1,\xi}^{n-1}, \quad \alpha_{k,\xi}^n = \frac{n}{\xi_{k+n} - \xi_k},$$

where terms with denominator zero are discarded.

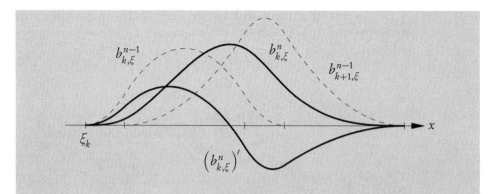

It follows from the recursion that $b_{k,\xi}^n$ is $(n-\mu)$-times continuously differentiable at a knot ξ_ℓ if ξ_ℓ has multiplicity $\mu \leq n$ among $\xi_k, \ldots, \xi_{k+n+1}$. In particular, $b_{k,\xi}^n$ is continuous on $\mathbb{R}$ unless one of its knots has multiplicity $n+1$.

We derive the formula from the recursion for B-splines,

$$b_{k,\xi}^n = \gamma_{k,\xi}^n b_{k,\xi}^{n-1} + (1-\gamma_{k+1,\xi}^n)b_{k+1,\xi}^{n-1}, \quad \gamma_{k,\xi}^n(x) = \frac{x-\xi_k}{\xi_{k+n}-\xi_k},$$

using induction on the degree n. There are several more elegant and less technical proofs. However, they do rely on further results about B-splines, e.g., their relationship to divided differences. The arguments below, based on the algorithmic approach by de Boor and Höllig, show in fact that the recursions for evaluation and differentiation are equivalent.

For the piecewise linear hat-functions $b_{k,\xi}^1$, the formula is easily checked. Hence, we turn to the induction step $n-1 \to n$. To this end, we note that

$$\alpha_{k,\xi}^n = n\left(\gamma_{k,\xi}^n\right)'$$

and set, in order to simplify notation,

$$B_k = b_{k,\xi}^{n-1}, \quad C_k = b_{k,\xi}^{n-2}, \quad \gamma_k = \gamma_{k,\xi}^n, \quad \delta_k = \gamma_{k,\xi}^{n-1}.$$

With these conventions, we write the assertion with the aid of the recursion in the equivalent form

$$\left(\gamma_k B_k + (1-\gamma_{k+1})B_{k+1}\right)' = n\gamma_k' B_k - n\gamma_{k+1}' B_{k+1}.$$

Using the induction hypothesis to differentiate B_k and B_{k+1}, the left side equals

$$\gamma_k' B_k - \gamma_{k+1}' B_{k+1} + (n-1)\Big[\gamma_k \delta_k' C_k - \gamma_k \delta_{k+1}' C_{k+1}$$
$$+ (1-\gamma_{k+1})\delta_{k+1}' C_{k+1} - (1-\gamma_{k+1})\delta_{k+2}' C_{k+2}\Big].$$

Comparing with the right side, it remains to show that the expression in brackets equals the right side without the factor n:

$$[\ldots] = \gamma_k' B_k - \gamma_{k+1}' B_{k+1}.$$

4.2. Differentiation

We now use the B-spline recursion a second time to express B_k and B_{k+1} as combinations of B-splines of degree $n-2$. This leads to

$$[\ldots] = \gamma'_k \delta_k C_k + \gamma'_k(1-\delta_{k+1})C_{k+1} - \gamma'_{k+1}\delta_{k+1}C_{k+1} - \gamma'_{k+1}(1-\delta_{k+2})C_{k+2}.$$

The validity of this equation can be checked by comparing the coefficients of the three B-splines C_k, C_{k+1}, and C_{k+2} on both sides of the identity. For example, the coefficients of C_k are

$$\text{left:} \quad \gamma_k \delta'_k = \frac{x-\xi_k}{\xi_{k+n}-\xi_k} \frac{d}{dx}\left(\frac{x-\xi_k}{\xi_{k+n-1}-\xi_k}\right);$$

$$\text{right:} \quad \gamma'_k \delta_k = \frac{d}{dx}\left(\frac{x-\xi_k}{\xi_{k+n}-\xi_k}\right) \frac{x-\xi_k}{\xi_{k+n-1}-\xi_k}.$$

The equality of these expressions is immediately apparent. Similarly, the equality of the coefficients of the other two B-splines can be verified by direct computation. For C_{k+1}, this is a bit tedious; the assistance of computer algebra is helpful.

The statement about the smoothness at the knots also follows by induction since a knot ξ_ℓ with multiplicity μ among $\xi_k, \ldots, \xi_{k+n+1}$ has at most the same multiplicity among the knots of the B-splines $b^{n-1}_{k,\xi}$ and $b^{n-1}_{k+1,\xi}$ appearing in the formula for $(b^n_{k,\xi})'$. Hence, since the B-splines of degree $n-1$ are at least $(n-1)-\mu$ times continuously differentiable at ξ_ℓ, by induction hypothesis, $b^n_{k,\xi}$ has one more continuous derivative.

The above argument is not applicable at a knot with multiplicity $\mu \geq n$. Without knowing that the B-splines of degree $n-1$ are continuous, which implies continuity of $b^n_{k,\xi}$ by the recursive definition, the formula for the derivatives of the polynomial segments need not represent the global derivative in a neighborhood of a discontinuity.

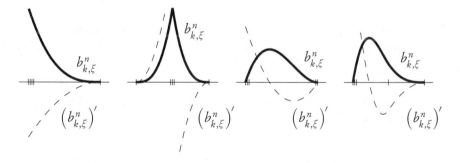

Fortunately, up to symmetry, there are only four possibilities, as shown in the figure.

In the first case, $\mu = n+1$ for ξ_k, and no continuity is asserted at this knot. In the second case, continuity at $\xi_{k+1} = \cdots = \xi_{k+n}$ follows from the explicit formulas for the first and last B-spline segments. In the remaining two cases, continuity at $\xi_k = \cdots = \xi_{k+n-1}$ is a consequence of the recursion:

$$b^n_{k,\xi}(x) = \frac{x-\xi_k}{\xi_{k+n}-\xi_k} b^{n-1}_{k,\xi}(x) + \frac{\xi_{k+n+1}-x}{\xi_{k+n+1}-\xi_{k+1}} b^{n-1}_{k+1,\xi}(x).$$

The B-spline $b^{n-1}_{k+1,\xi}$ is continuous by induction, and the zero of the linear factor $\gamma^n_{k,\xi}$ at ξ_k ensures continuity also of the first term.

☐ **Example:**
The figure shows two cubic B-splines together with their first and second derivatives on the knot intervals.

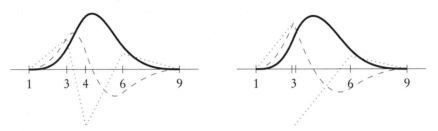

The B-spline on the left has simple knots. Accordingly, it is twice continuously differentiable ($n - \mu = 3 - 1 = 2$). The B-spline on the right has simple knots at $x = 1, 6, 9$ and a double knot (multiplicity $\mu = 2$) at 3. Hence, it is twice continuously differentiable, except at the double knot where only the first derivative is continuous ($n - \mu = 3 - 2 = 1$). Differentiating $b_{0,\xi}^3$, the recursion yields

$$(b_{0,\xi}^3)' = \frac{3}{6-1} b_{0,\xi}^2 - \frac{3}{9-3} b_{1,\xi}^2 = \frac{3}{5} b_{0,\xi}^2 - \frac{1}{2} b_{1,\xi}^2 .$$

Applying the recursion to both B-splines on the right side,

$$(b_{0,\xi}^3)'' = \frac{3}{5}\left(\frac{2}{3-1} b_{0,\xi}^1 - \frac{2}{6-3} b_{1,\xi}^1\right) - \frac{1}{2}\left(\frac{2}{6-3} b_{1,\xi}^1 - \frac{2}{9-3} b_{2,\xi}^1\right) = \frac{3}{5} b_{0,\xi}^1 - \frac{11}{15} b_{1,\xi}^1 + \frac{1}{6} b_{2,\xi}^1 .$$

We see that the second derivative is discontinuous at $x = 3$:

$$(b_{0,\xi}^3)''(3^-) = 3/5, \quad (b_{0,\xi}^3)''(3^+) = -11/15 .$$

Hence, a global third derivative does not exist. Nevertheless, we can differentiate separately on each knot interval. For $x \in [1,3), [3,6), [6,9)$, the recursion yields the values $3/10, 3/10, -1/18$, respectively. ☐

With the aid of the differentiation formula, we can also determine the precise behavior of a B-spline at the endpoints of its support.

Zeros of B-Splines
If ξ_k has multiplicity μ among the knots of $b_{k,\xi}^n$, then the left endpoint ξ_k of the B-spline support is a zero of order $n + 1 - \mu$:

$$b_{k,\xi}^n(x) = \beta_k (x - \xi_k)^{n+1-\mu} + \mathcal{O}\left((x - \xi_k)^{n+2-\mu}\right), \quad x \to \xi_k^+,$$

with $\beta_k > 0$.
An analogous statement is valid for the right endpoint of the support of $b_{k,\xi}^n$.

This statement follows by induction on the degree. For the induction step from $n-1$ to n, we just have to note that the two B-splines of degree $n-1$ in the expression for $(b_{k,\xi}^n)'$

have zeros of order $n-\mu$ and $n-(\mu-1)$, respectively. Hence, $(b_{k,\xi}^n)'$ has a zero of order $n-\mu$ at ξ_k. Integration increases the order by 1. The positivity of β_k is a consequence of the positivity of the B-spline.

4.3 ▪ Representation of Polynomials

While a B-spline consists of polynomial segments of degree $\leq n$, it is by no means clear that we can represent polynomials as linear combinations of B-splines. Clearly, this is necessary in order to maintain the accuracy of polynomial approximations. Hence, the following formula is a crucial identity in spline theory.

> **Marsden's Identity**
> For a bi-infinite knot sequence ξ, any polynomial of degree $\leq n$ can be represented as a linear combination of B-splines. In particular, for any $y \in \mathbb{R}$,
> $$(x-y)^n = \sum_{k\in\mathbb{Z}} \psi_{k,\xi}^n(y) b_{k,\xi}^n(x), \ x \in \mathbb{R},$$
> with $\psi_{k,\xi}^n(y) = (\xi_{k+1}-y)\cdots(\xi_{k+n}-y)$.
>
>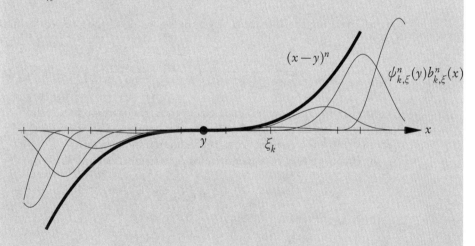
>
> Comparing coefficients of y^{n-m} on both sides of the identity yields explicit representations for the monomials x^m. In particular, we have
> $$1 = \sum_k b_{k,\xi}^n(x), \quad x = \sum_k \xi_k^n \, b_{k,\xi}^n(x)$$
> with $\xi_k^n = (\xi_{k+1}+\cdots+\xi_{k+n})/n$ the so-called knot averages or Greville abscissae.

To derive Marsden's formula, we use induction on the degree n. Clearly, for $n=0$,
$$(x-y)^0 = 1 = \sum_k b_{k,\xi}^0(x),$$
noting that the empty product $\psi_{k,\xi}^0(y)$ is equal to 1 and, by assumption, the knot intervals cover the entire real line.

For the induction step from $n-1$ to n we rewrite the assertion with the aid of the B-spline recursion as

$$(x-y)^n = \sum_k \psi_k(y)(\gamma_k(x)b_{k,\xi}^{n-1}(x) + (1-\gamma_{k+1}(x))b_{k+1,\xi}^{n-1}(x)),$$

where $\gamma_k(x) = \gamma_{k,\xi}^n(x) = (x-\xi_k)/(\xi_{k+n}-\xi_k)$ and $\psi_k = \psi_{k,\xi}^n$. Shifting the index of the second summand ($k \to k-1$), the right side becomes

$$\sum_k \left[\gamma_k(x)\psi_k(y) + (1-\gamma_k(x))\psi_{k-1}(y)\right] b_{k,\xi}^{n-1}(x).$$

Since the functions ψ_k and ψ_{k-1} contain the common factor

$$\psi_{k,\xi}^{n-1}(y) = (\xi_{k+1}-y)\cdots(\xi_{k+n-1}-y),$$

the expression in brackets simplifies:

$$[\ldots] = \left(\gamma_k(x)(\xi_{k+n}-y) + (1-\gamma_k(x))(\xi_k-y)\right)\psi_{k,\xi}^{n-1}(y).$$

Hence, recalling the definition of γ_k, the right side of the asserted formula for $(x-y)^n$ equals

$$\sum_k \left(\frac{(x-\xi_k)(\xi_{k+n}-y)}{\xi_{k+n}-\xi_k} + \frac{(\xi_{k+n}-x)(\xi_k-y)}{\xi_{k+n}-\xi_k}\right)\psi_{k,\xi}^{n-1}(y)b_{k,\xi}^{n-1}(x).$$

A straightforward computation, e.g., expanding with respect to x, shows that the sum in parentheses equals $(x-y)$. Cancelling this common factor on both sides of Marsden's identity shows the equivalence to the induction hypothesis.

It remains to verify the identities for the monomials 1 and x. To this end, we note that

$$(x-y)^n = (-1)^n y^n + n(-1)^{n-1}xy^{n-1} + \cdots$$

and

$$\psi_k(y) = (-1)^n y^n + (-1)^{n-1}(\xi_{k+1} + \cdots + \xi_{k+n})y^{n-1} + \cdots,$$

in each case neglecting terms of order $O(y^{n-2})$. By comparing the coefficients of the two highest powers of y we obtain the asserted formulas.

□ **Example:**

We determine the representation of the monomial x^2. Comparing the coefficients of y^{n-2} on both sides of Marsden's identity yields, after division by $(-1)^{n-2}\binom{n}{2}$,

$$x^2 = \sum_k c_k\, b_k(x), \quad c_k = \frac{2}{n(n-1)}\sum_{1 \le i < j \le n} \xi_{k+i}\xi_{k+j}$$

since the sum equals the coefficient of $(-y)^{n-2}$ in an expansion of $\psi_{k,\xi}^n(y)$.

We can write the formula for the coefficients c_k in a more compact form. With

$$\xi_k^n = \frac{1}{n}\sum_{\ell=1}^{n}\xi_{k+\ell}, \quad \sigma_k^2 = \frac{1}{n-1}\sum_{\ell=1}^{n}(\xi_{k+\ell}-\xi_k^n)^2$$

denoting the knot averages and the geometric variance of n consecutive knots,

$$c_k = (\xi_k^n)^2 - \sigma_k^2/n.$$

To prove the alternative formula, we compare the coefficients $\gamma_{i,j}$ of the terms $\xi_{k+i}\xi_{k+j}$ in the two expressions for the coefficients c_k. For the first expression,

$$\gamma_{i,i}^{I} = 0, \quad \gamma_{i,j}^{I} = \frac{2}{n(n-1)}, \quad i < j.$$

The second expression equals

$$\frac{1}{n^2}(\xi_{k+1}+\cdots+\xi_{k+n})^2 - \frac{1}{n^3(n-1)}\sum_{\ell=1}^{n}(n\xi_{k+\ell}-\xi_{k+1}-\cdots-\xi_{k+n})^2$$

in view of the definitions of ξ_k^n and σ_k^2. Expanding both squares $(...)^2$ with the binomial formula, we obtain

$$\gamma_{i,i}^{II} = \frac{1}{n^2} - \frac{1}{n^3(n-1)}((n-1)+(n-1)^2)$$

and, for $i < j$,

$$\gamma_{i,j}^{II} = \frac{2}{n^2} - \frac{1}{n^3(n-1)}(2(n-2)-2\cdot 2(n-1)).$$

After simplification, $\gamma^I = \gamma^{II}$, as desired. $\qquad\square$

Marsden's identity was formulated for a bi-infinite knot sequence which extends over the entire real line. Obviously, we can consider finite knot sequences as well. Restricting x to a knot interval $D_\ell = [\xi_\ell, \xi_{\ell+1})$, only B-splines with supports overlapping D_ℓ are relevant:

$$(x-y)^n = \sum_{k=\ell-n}^{\ell} \psi_{k,\xi}^n(y)b_{k,\xi}^n(x), \quad \xi_\ell \le x < \xi_{\ell+1},$$

provided that ξ contains the knots $\xi_{\ell-n}\,...,\xi_{\ell+n+1}$ involved.

4.4 ▪ Splines

The classical definition of splines imposes smoothness constraints on piecewise polynomials on a partition of a parameter interval. B-splines resolve the resulting coupling of the polynomial segments in a very elegant fashion, laying the foundation to the modern treatment of splines. Obviously, the local support of the B-spline basis has numerical advantages. Moreover, B-splines facilitate the theoretical analysis of spline spaces.

To define spline spaces, some notation is useful.

Parameter Interval and Regularity of Knot Sequences

The parameter interval D_ξ^n is the maximal interval on which the B-splines $b_{k,\xi}^n$, $k \sim \xi$, which correspond to the knot sequence ξ, form a partition of unity.

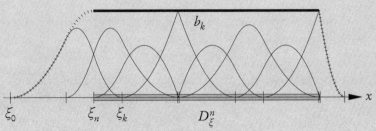

We say that a knot sequence ξ is n-regular if each B-spline b_k, $k \sim \xi$, is continuous and nonzero at some points in D_ξ^n. More explicitly, n-regularity requires that all knot multiplicities are $\leq n$ and, for a finite knot sequence $\xi : \xi_0, \ldots, \xi_{m+n}$ in addition, that

$$\xi_n < \xi_{n+1}, \quad \xi_{m-1} < \xi_m.$$

Let us comment on these definitions in some more detail. For a bi-infinite knot sequence, $D_\xi^n = \mathbb{R}$, and, for a finite knot sequence,

$$\xi_0, \ldots, \xi_{m+n},$$

$D_\xi^n = [\xi_n, \xi_m]$ unless $\xi_m = \cdots = \xi_{m+n}$ is a knot of multiplicity $n + 1$, in which case $D_\xi^n = [\xi_n, \xi_m)$. The distinction is necessary since in the latter case the B-splines $b_{k,\xi}^n$, $k \sim \xi$, vanish at ξ_m due to the convention of continuity from the right. Assuming n-regularity avoids such technical difficulties and other pathological cases which are of no relevance in practice. We give two examples for degree $n = 2$.

(i) Discontinuous B-splines: For the knot sequence

$$\xi_0 = 0, 0, 0, 1, 1, 1 = \xi_5,$$

all B-splines $b_{k,\xi}^2$, $k = 0, 1, 2$, vanish at $x = 1$. This implies that $D_\xi^2 = [0, 1)$, which is not a suitable choice for approximation of functions or parametrization of curves. Extending the knot sequence with $\xi_6 = 2$, the situation becomes worse. Now, $D_\xi^2 = [0, 1]$, but the B-spline $b_{3,\xi}^2$ contributes only a value at a single point. Of course, we could alter the definition of B-splines at the interval endpoint $x = 1$. But working with different types of B-splines has obvious drawbacks.

(ii) B-splines without support in the parameter interval: For the knot sequence

$$\xi_0 = -1, -1, 0, 0, 1, 1, 2, 2 = \xi_7,$$

$D_\xi^2 = [0, 1]$, but the B-splines $b_{0,\xi}^2$ and $b_{4,\xi}^2$ are nonzero only at points outside of the parameter interval. The inequalities $\xi_n < \xi_{n+1}$ and $\xi_{m-1} < \xi_m$, mentioned above, exclude this possibility.

For both of the above examples, the knot sequence

$$\xi_0 = -1, 0, 0, 1, 1, 2 = \xi_5$$

4.4. Splines

is a better choice. In this case, the B-splines $b^2_{k,\xi}$, $k \sim \xi$, coincide with the Bernstein polynomials on the closed parameter interval $D^2_\xi = [0,1]$:

$$b^2_{k,\xi}(x) = b^2_k(x), \quad x \in [0,1].$$

Reiterating a point made above, the knot sequence ξ in the example (i) does not lead to the Bernstein basis since $b^2_{2,\xi}(1) = 0 \neq 1 = b^2_2(1)$ for a triple knot at 1.

Using n-regular knot sequences, we avoid the problems inherent in the assumption of continuity from the right. In particular, we can work with closed knot intervals $[\xi_\ell, \xi_{\ell+1}]$, disposing with the asymmetry of the half-open knot intervals. Although n-regularity implies that the degree n is greater than 0, this is no major drawback.

After these preliminaries, we can now define splines in a familiar fashion.

Splines

A spline p of degree $\leq n$ with $n > 0$ is a linear combination of the B-splines corresponding to an n-regular knot sequence ξ:

$$p(x) = \sum_k c_k b^n_{k,\xi}(x), \quad x \in D^n_\xi.$$

The coefficients are unique, i.e., the B-splines b_k, $k \sim \xi$, restricted to D^n_ξ form a basis for the spline space denoted by S^n_ξ.

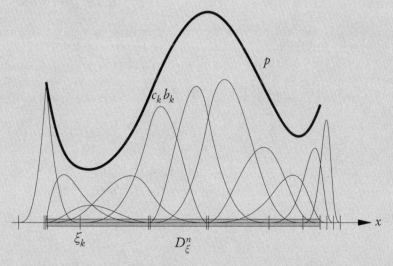

Equivalently, S^n_ξ consists of all continuous functions on the parameter interval D^n_ξ which are

- polynomials of degree $\leq n$ on the nondegenerate knot intervals of D^n_ξ;
- $n - \mu$ times continuously differentiable at an interior knot of D^n_ξ with multiplicity μ.

We first prove the linear independence of the B-splines b_k, $k \sim \xi$, on D_ξ^n. To this end, we assume that

$$p(x) = \sum_k c_k \, b_k(x) = 0 \quad \forall x \in D_\xi^n$$

and show that this implies $c_k = 0$ for all k. The argument is based on Marsden's identity. On a nondegenerate interval $D_\ell = [\xi_\ell, \xi_{\ell+1}] \subseteq D_\xi^n$, the linear span of the relevant B-splines

$$b_{\ell-n}, \ldots, b_\ell$$

contains all polynomials of degree $\leq n$ on D_ℓ, a vector space of dimension $n+1$. Hence, they are linearly independent on D_ℓ, and, consequently, $c_{\ell-n} = \cdots = c_\ell = 0$. Considering all possible intervals D_ℓ, we conclude that $c_k = 0$ for all $k \sim \xi$ since all B-splines b_k are nonzero at some points in D_ξ^n.

We now derive the alternative classical description of the spline space. To this end, we denote the space of piecewise polynomials of degree $\leq n$ constrained by the smoothness conditions by P_ξ^n. Clearly,

$$S_\xi^n \subseteq P_\xi^n$$

since any of the B-splines b_k meets the smoothness requirements.

For the reverse inclusion, we have to express a piecewise polynomial $p \in P_\xi^n$ as a linear combination of B-splines. We begin by considering a nondegenerate knot interval $D_\ell = [\xi_\ell, \xi_{\ell+1}] \subseteq D_\xi^n$. Since p is a polynomial of degree $\leq n$ on this interval, there exist coefficients c_k with

$$p(x) = q_\ell(x) = \sum_{k=\ell-n}^{\ell} c_k b_k(x), \quad x \in D_\ell,$$

by Marsden's identity. We now extend this representation to the neighboring knot intervals. Proceeding, e.g., to the right, we assume that $\xi_{\ell+1}$ is an interior knot with multiplicity μ (otherwise the representation is already valid up to the right boundary of D_ξ^n). Then, p as well as q_ℓ are $n - \mu$ times continuously differentiable at $\xi_{\ell+1}$. Hence, the derivatives of $p - q_\ell$ vanish at $\xi_{\ell+1}$ up to order $n - \mu$. It follows that

$$p(x) - q_\ell(x) = \sum_{\alpha=1}^{\mu} d_\alpha (x - \xi_{\ell+1})^{n-\mu+\alpha}$$

for x in the neighboring nondegenerate knot interval $D_{\ell+\mu}$. This sum of μ monomials can be expressed as a linear combination $q_{\ell+\mu}$ of the μ B-splines $b_{\ell+1}, \ldots, b_{\ell+\mu}$ since

$$b_{\ell+\alpha} = \beta_{\ell+\alpha}(x - \xi_{\ell+1})^{n-\mu+\alpha} + O\big((x - \xi_{\ell+1})^{n-\mu+\alpha+1}\big), \quad \beta_{\ell+\alpha} > 0$$

(cf. the statement at the end of Section 4.2). Clearly, the identity

$$p(x) - q_\ell(x) = q_{\ell+\mu}(x), \quad x \in D_{\ell+\mu},$$

is valid also on D_ℓ, where both sides are zero. Hence, we have expressed p in terms of the B-splines b_k, $k \sim \xi$, on two adjacent nondegenerate knot intervals ($p = q_\ell + q_{\ell+\mu}$). Proceeding in this fashion to the right as well as to the left, we eventually exhaust the entire parameter interval D_ξ^n. The argument applies to a finite as well as a bi-infinite knot sequence. In the latter case, $D_\xi^n = \mathbb{R}$ and infinitely many steps are required in both directions.

4.4. Splines

☐ **Example:**

In applications, cubic splines are particularly important. Below we consider two frequently used knot sequences.

(i) Simple knots: If
$$\xi_0 < \cdots < \xi_{3+m},$$
the spline space S_ξ^3 consists of all twice continuously differentiable functions which are polynomials of degree ≤ 3 on each of the knot intervals $[\xi_\ell, \xi_{\ell+1}]$ in $D_\xi^3 = [\xi_3, \xi_m]$.

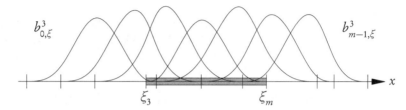

(ii) Double knots: If
$$\xi_0 = \xi_1 < \xi_2 = \xi_3 < \cdots < \xi_m = \xi_{m+1} < \xi_{m+2} = \xi_{m+3},$$
the second derivatives of the cubic splines can have jumps. In this case, on each knot interval $[\xi_{2k-1}, \xi_{2k}]$, the cubic splines are uniquely determined by their values and derivatives at the knots. These data provide an alternative to the B-spline basis for specifying a spline in S_ξ^3.

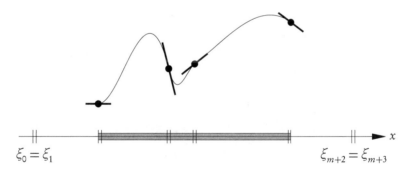

☐

For a knot sequence $\xi_0, \ldots, \xi_{n+m}$, the B-spline basis for the spline space S_ξ^n requires n exterior knots on each side of the parameter interval $D_\xi^n = [\xi_n, \xi_m]$. These knots ξ_k with $k < n$ or $k > m$ are irrelevant for the definition of S_ξ^n in terms of its piecewise polynomial structure. They affect, however, the B-spline basis. There are two standard choices.

(i) Simple exterior knots: With $\Delta \xi_\ell = \xi_{\ell+1} - \xi_\ell$, one defines
$$\xi_{n-\ell} = \xi_n - \ell \Delta \xi_n, \quad \xi_{m+\ell} = \xi_m + \ell \Delta \xi_{m-1}$$
for $\ell = 1, \ldots, n$. This choice keeps the knot spacing of the first and last knot intervals in D_ξ^n and maximizes the smoothness of the B-splines.

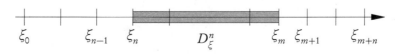

(ii) Multiple boundary knots: One defines

$$\xi_1 = \cdots = \xi_n, \quad \xi_m = \cdots = \xi_{m+n-1}$$

and chooses only ξ_0 and ξ_{m+n} outside of the interval D_ξ^n. Because of the maximal admissible multiplicity, only the B-splines b_0 and b_{m-1} are nonzero at the interval endpoints and are equal to 1 at these points. Hence, for a spline $p = \sum_{k=0}^{m-1} c_k b_k$,

$$p(\xi_n) = c_0, \quad p(\xi_m) = c_{m-1}.$$

The drawback of knot multiplicity n is that the derivatives of the B-splines are no longer continuous. Thus, the convention of continuity from the right leads to an asymmetric treatment of the endpoints of D_ξ^n.

When approximating with B-splines, the knot spacing can be adapted to the local complexity of the data. However, this flexibility is not needed often, and a uniform length of the knot intervals is chosen. This important special case, extensively studied by Schoenberg, will now be discussed in more detail.

Uniform B-Splines

The standard uniform B-spline b^n has the knots $0, 1, \ldots, n+1$. In this special case, the recursions for evaluation and differentiation simplify:

$$n b^n(x) = x b^{n-1}(x) + (n+1-x) b^{n-1}(x-1),$$
$$\frac{d}{dx} b^n(x) = b^{n-1}(x) - b^{n-1}(x-1).$$

Moreover, the second identity can be written as an averaging process:

$$b^n(x) = \int_0^1 b^{n-1}(x-y)\, dy.$$

The uniform B-splines for an arbitrary uniform knot sequence with grid width h, i.e., with knot intervals $\xi_\ell + [0, h]$, are scaled translates of b^n:

$$b_{k,\xi}^n(x) = b^n((x - \xi_k)/h), \quad k \sim \xi.$$

For a bi-infinite uniform knot sequence $\xi_k = kh$ is the standard choice. However, for a finite uniform knot sequence, a shift of the index might be convenient and necessary. For example, the uniform knot sequence for splines S_ξ^n with parameter interval $D_\xi^n = [0,1]$ is

$$-nh,\ldots,-h,0,h,\ldots,1-h,1,1+h,\ldots,1+nh.$$

Hence, $\xi_k = (k-n)h$ leads to the standard range $\{0,\ldots,m-1\}$ of relevant indices k.

The recursions for general B-splines are easily specialized to the uniform case. Therefore, it remains to verify the averaging formula. To this end, we integrate the recursion for the derivative and obtain

$$b^n(x) = \int_{-\infty}^{x} b^{n-1}(z)\,dz - \int_{-\infty}^{x-1} b^{n-1}(z)\,dz.$$

Combining the integrals gives

$$\int_{x-1}^{x} b^{n-1}(z)\,dz,$$

which, after the substitution $z = x - y$, is of the desired form.

□ **Example:**

If we denote the values of the nth derivative of b^n (which is piecewise constant) on the knot intervals by $d_0^n,\ldots,d_n^n$, it follows from the recursion for the derivatives that

$$d^n = (d^{n-1},0) - (0,d^{n-1}).$$

This yields

$$d^0 : 1,$$
$$d^1 : 1,-1,$$
$$d^2 : 1,-2,1,$$
$$d^3 : 1,-3,3,-1,$$

and, in general,

$$d_k^n = (-1)^k \binom{n}{k}, \quad k = 0,\ldots,n. \qquad \square$$

We can formulate the recursion for evaluation also as a recursion for the Taylor coefficients of the polynomial segments of a standard uniform B-spline. For $x \in [k,k+1)$, we define

$$p_k^n(y) = \sum_{\ell=0}^{n} a_{k,\ell}^n y^\ell = b^n(x), \quad x - k = y \in [0,1).$$

Then, the recursion can be rewritten as

$$n p_k^n(y) = (k+y)p_k^{n-1}(y) + (n+1-k-y)p_{k-1}^{n-1}(y), \quad k = 0,\ldots,n,$$

with $p_{-1}^{n-1} = p_n^{n-1} = 0$. The corresponding identity for the coefficients is

$$na_{k,\ell}^n = ka_{k,\ell}^{n-1} + a_{k,\ell-1}^{n-1} + (n+1-k)a_{k-1,\ell}^{n-1} - a_{k-1,\ell-1}^{n-1},$$

where $a_{k,-1}^{n-1} = a_{k,n}^{n-1} = a_{-1,\ell}^{n-1} = a_{n,\ell}^{n-1} = 0$.

The first few coefficient vectors are tabulated below.

n	a_0^n	a_1^n	a_2^n	a_3^n
1	$(0, 1)$	$(1, -1)$		
2	$\left(0, 0, \frac{1}{2}\right)$	$\left(\frac{1}{2}, 1, -1\right)$	$\left(\frac{1}{2}, -1, \frac{1}{2}\right)$	
3	$\left(0, 0, 0, \frac{1}{6}\right)$	$\left(\frac{1}{6}, \frac{1}{2}, \frac{1}{2}, -\frac{1}{2}\right)$	$\left(\frac{2}{3}, 0, -1, \frac{1}{2}\right)$	$\left(\frac{1}{6}, -\frac{1}{2}, \frac{1}{2}, -\frac{1}{6}\right)$

With the aid of the tabulated Taylor coefficients we can evaluate standard uniform B-splines faster than with the recursion. For example, for $x \in [3,4) \Leftrightarrow k = 3 \wedge y \in [0,1)$,

$$b^3(x) = p_3^3(y) = \frac{1}{6} - \frac{1}{2}y + \frac{1}{2}y^2 - \frac{1}{6}y^3 = (((-y+3)y-3)y+1)/6.$$

4.5 ▪ Evaluation and Differentiation

The recursions for B-splines yield analogous formulas for evaluation and differentiation of splines. We discuss each operation in turn.

De Boor Algorithm

A linear combination of B-splines

$$p = \sum_k c_k b_k$$

of degree $\leq n$ with knot sequence ξ can be evaluated at $x \in [\xi_\ell, \xi_{\ell+1})$ by forming convex combinations of the coefficients of the B-splines which are nonzero at x. Starting with

$$p_k^0 = c_k, \quad k = \ell - n, \ldots, \ell,$$

one computes successively, for $i = 0, \ldots, n-1$,

$$p_k^{i+1} = \gamma_{k,\xi}^{n-i} p_k^i + (1 - \gamma_{k,\xi}^{n-i})p_{k-1}^i, \quad k = \ell - n + i + 1, \ldots, \ell,$$

with

$$\gamma_{k,\xi}^{n-i} = \frac{x - \xi_k}{\xi_{k+n-i} - \xi_k}$$

and obtains $p(x)$ as the final value p_ℓ^n.

4.5. Evaluation and Differentiation

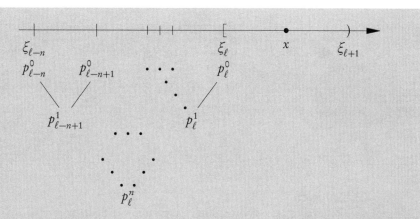

The triangular scheme is illustrated in the figure. It simplifies slightly if $x = \xi_\ell$. If ξ_ℓ has multiplicity μ, then $p(x) = p_{\ell-\mu}^{n-\mu}$; i.e., only $n - \mu$ steps of the recursion applied to $c_{\ell-n}, \ldots, c_{\ell-\mu}$ are needed.

The algorithm is based on the recursion for B-splines:

$$b_k = b_{k,\xi}^n = \gamma_{k,\xi}^n b_{k,\xi}^{n-1} + (1 - \gamma_{k+1,\xi}^n) b_{k+1,\xi}^{n-1}.$$

Substituting this identity into

$$p(x) = \sum_k p_k^0 b_k(x),$$

fixing x, and shifting the index in the second term, we obtain

$$p = \sum_k \left[\gamma_{k,\xi}^n p_k^0 + (1 - \gamma_{k,\xi}^n) p_{k-1}^0 \right] b_{k,\xi}^{n-1}.$$

We now abbreviate the expression in brackets by p_k^1, and apply the B-spline recursion once more to rewrite $b_{k,\xi}^{n-1}$ in terms of B-splines of degree $n-2$. Again, convex combinations of the coefficients are formed:

$$p_k^2 = \gamma_{k,\xi}^{n-1} p_k^1 + (1 - \gamma_{k,\xi}^{n-1}) p_{k-1}^1.$$

Continuing in this fashion, we finally arrive at

$$p = \sum_k p_k^n b_{k,\xi}^0.$$

Since the B-splines $b_{k,\xi}^0$ are the characteristic functions of the knot intervals and $x \in [\xi_\ell, \xi_{\ell+1})$, it follows that $p(x) = p_\ell^n$. It remains to note that for the final value only the coefficients

$$c_{\ell-n}, \ldots, c_\ell$$

are relevant, i.e., the coefficients

$$p_k^i, \quad k = \ell - n + i, \ldots, \ell,$$

at stage i of the triangular scheme.

If
$$\xi_{\ell-\mu} < \xi_{\ell-\mu+1} = \cdots = \xi_\ell = x < \xi_{\ell+1},$$
some of the relevant B-splines for the knot interval $[\xi_\ell, \xi_{\ell+1})$ vanish at x. As a consequence,
$$\gamma^1_{\ell,\xi} = \gamma^2_{\ell-1,\xi} = \cdots = \gamma^\mu_{\ell-\mu+1,\xi} = 0,$$
and, hence,
$$p^n_\ell = p^{n-1}_{\ell-1} = \cdots = p^{n-\mu}_{\ell-\mu},$$
as claimed.

☐ **Example:**
We evaluate the cubic spline p shown in the figure at $x = 5$.

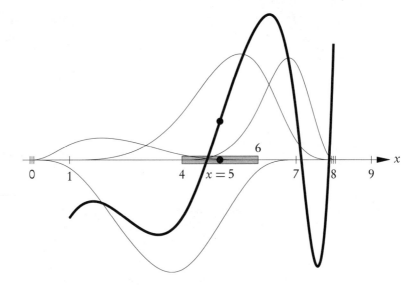

The coefficients and knots in this example are
$$c_0 = -3, 1, -5, 4, 4, -7, 9 = c_6, \quad \xi_0 = 0, 0, 0, 1, 4, 6, 7, 8, 8, 8, 9 = \xi_{10}.$$
The parameter x lies in the knot interval $[\xi_4, \xi_5) = [4,6)$ for which the coefficients
$$c_1 = 1, c_2 = -5, c_3 = 4, c_4 = 4$$
are relevant. Applying the triangular scheme, as shown below, yields $p(x) = p^3_4 = 1$.

4.5. Evaluation and Differentiation

If we evaluate p at $x = 4 = \xi_4$, only two steps of the triangular scheme are needed. The value of the spline depends only on c_1, c_2, c_3: $p(4) = p_3^2 = -3/2$.

```
c₁ = 1
            1/3
      2/3
c₂ = -5           p₂¹ = -3
      1/2                    2/5
      1/2           3/5
c₃ = 4            p₃¹ = -1/2           p₃² = -3/2
```

When implementing an algorithm for evaluation, a special treatment of evaluation at knots does not save much time. However, a problem can arise if not enough B-spline coefficients are available for the full n-stage triangular scheme. This difficulty can be most easily resolved by extending the knot sequence and setting the auxiliary B-spline coefficients to zero. □

We now turn to differentiation of splines. The formula for the derivative given below is typical for the B-spline calculus. Operations on splines correspond to analogous operations on the B-spline coefficients.

Differentiation

The derivative of a spline is a linear combination of B-splines with the same knot sequence. More precisely, for x in any open knot interval $(\xi_\ell, \xi_{\ell+1})$ with $n+1$ relevant B-splines $b_{k,\xi}^n$,

$$\frac{d}{dx}\left(\sum_{k=\ell-n}^{\ell} c_k b_{k,\xi}^n(x)\right) = \sum_{k=\ell-n+1}^{\ell} \alpha_{k,\xi}^n \nabla c_k \, b_{k,\xi}^{n-1}(x), \quad \alpha_{k,\xi}^n = \frac{n}{\xi_{k+n} - \xi_k},$$

with ∇ the backward difference operator, i.e., $\nabla c_k = c_k - c_{k-1}$. The identity remains valid at the endpoints ξ_ℓ and $\xi_{\ell+1}$ of the knot interval if these knots have multiplicity $< n$.

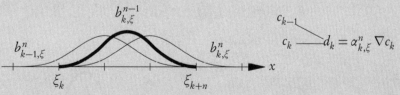

For a spline $p = \sum_{k=0}^{m-1} c_k b_k \in S_\xi^n$ with knot sequence $\xi_0, \ldots, \xi_{m+n}$ and multiplicities $< n$,

$$p' = \sum_{k=1}^{m-1} d_k b_{k,\xi}^{n-1} \in S_{\xi'}^{n-1},$$

where ξ' is obtained from ξ by deleting the first and last knots. This is consistent with the difference operation ∇, which reduces the range of indices by 1.

We recall that a spline is assumed to be continuous. Since differentiation can cause discontinuities, the derivative of a spline is, in general, merely a linear combination of

B-splines. Continuity is maintained only if the knot multiplicities are less than the degree n.

The formula for the derivative is a direct consequence of the recursion for the derivative of a B-spline:
$$\left(b_{k,\xi}^n\right)' = \alpha_{k,\xi}^n b_{k,\xi}^{n-1} - \alpha_{k+1,\xi}^n b_{k+1,\xi}^{n-1}.$$

For a linear combination of B-splines on the knot interval $(\xi_\ell, \xi_{\ell+1})$, we obtain
$$\left(\sum_{k=\ell-n}^{\ell} c_k b_{k,\xi}^n\right)' = \sum_{k=\ell-n}^{\ell} c_k \left[\alpha_{k,\xi}^n b_{k,\xi}^{n-1} - \alpha_{k+1,\xi}^n b_{k+1,\xi}^{n-1}\right].$$

Shifting the index in the second term of the expression in brackets ($k \to k-1$), the right side equals
$$\sum_{k=\ell-n}^{\ell} c_k \alpha_{k,\xi}^n b_{k,\xi}^{n-1} - \sum_{k=\ell-n+1}^{\ell+1} c_{k-1} \alpha_{k,\xi}^n b_{k,\xi}^{n-1}.$$

Combining the two sums and noting that $b_{\ell-n,\xi}^{n-1}$ and $b_{\ell+1,\xi}^{n-1}$ vanish on $(\xi_\ell, \xi_{\ell+1})$ yields the desired form of the derivative.

If all knots $\xi_0, \ldots, \xi_{m+n}$ of a spline space S_ξ^n have multiplicity $< n$, the B-splines
$$b_0, \ldots, b_{m-1}$$
are continuously differentiable. Therefore, the formula for the derivative is valid across knots, i.e., the sum can be taken over all B-splines, relevant for the parameter interval $D_\xi^n = [\xi_n, \xi_m]$. For degree $n-1$, these are $b_{k,\xi}^{n-1}$ with $k = 1, \ldots, m-1$; $b_{0,\xi}^{n-1}$ vanishes on D_ξ^n. This means that the knot sequence can be shortened at both ends (the last knot of $b_{m-1,\xi}^{n-1}$ is ξ_{m+n-1}) in accordance with the formula for the derivative.

☐ **Example:**

We differentiate the cubic spline
$$p = 3b_{0,\xi}^3 - b_{1,\xi}^3 + 4b_{2,\xi}^3 - 6b_{3,\xi}^3 + 9b_{4,\xi}^3 + 5b_{5,\xi}^3 - 7b_{6,\xi}^3 \in S_\xi^3,$$

shown in the figure, on the parameter interval $D_\xi^3 = [\xi_3, \xi_7] = [1, 9]$.

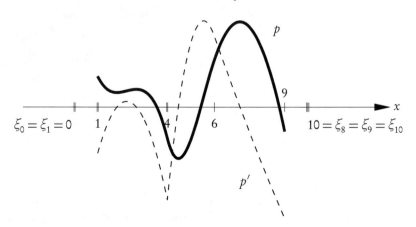

From the knot sequence

$$\xi_0 = 0, 0, 1, 1, 4, 4, 6, 9, 10, 10, 10 = \xi_{10},$$

we first compute the weights

$$\alpha_{k,\xi}^3 = 3/(\xi_{k+3} - \xi_k)$$

and obtain

$$\alpha_{1,\xi}^3 = 3/4, 1, 3/5, 3/5, 1/2, 3/4 = \alpha_{6,\xi}^3.$$

Then, we form the weighted differences $d_k = \alpha_k^3 \nabla c_k$ of the coefficients.

ξ_k	0	0	1	1	4	4	6	9	10	10	10
c_k	3	−1	4	−6	9	5	−7				
∇c_k		−4	5	−10	15	−4	−12				
α_k^3		3/4	1	3/5	3/5	1/2	3/4				
d_k		−3	5	−6	9	−2	−9				

Hence,

$$p' = -3b_{1,\xi}^2 + 5b_{2,\xi}^2 - 6b_{3,\xi}^2 + 9b_{4,\xi}^2 - 2b_{5,\xi}^2 - 9b_{6,\xi}^2$$

on D_ξ^n.

The differentiation formula simplifies considerably for uniform knots with grid width h. In this case, $\alpha_{k,\xi}^n = 1/h$, and, hence,

$$\left(\sum_k c_k \, b_{k,\xi}^n \right)' = h^{-1} \sum_k \nabla c_k \, b_{k,\xi}^{n-1};$$

i.e., differentiation just amounts to forming differences of the coefficients. $\square$

4.6 ▪ Periodic Splines

Periodic splines belong to subspaces of S_ξ^n for bi-infinite knot sequences. Hence, they do not require a special treatment. We just describe in the following how to characterize periodicity in terms of the knot sequence and the B-spline coefficients.

> **Periodic Splines**
>
> A spline
>
> $$p = \sum_{k \in \mathbb{Z}} c_k \, b_k \in S_\xi^n$$
>
> with bi-infinite knot sequence ξ is T-periodic if the knots ξ_k and the coefficients c_k satisfy the periodicity conditions
>
> $$\xi_{k+M} = \xi_k + T, \, c_{k+M} = c_k, \quad k \in \mathbb{Z},$$
>
> for some $M \in \mathbb{N}$.

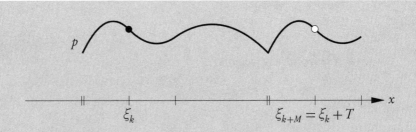

The periodic splines form a subspace $S^n_{\eta,T}$ of S^n_ξ of dimension M, where η is any subsequence of M consecutive knots of ξ.

The periodicity conditions are easily verified. Translating the knot vector of a B-spline is equivalent to the corresponding shift of the variable. Hence, it is obvious that splines for which the knot sequence and the coefficients satisfy the periodicity conditions are T-periodic. Conversely, it is plausible that all periodic splines are of this form. Rigorously, this can be proved as follows.

Clearly, periodicity implies that the smoothness conditions at all knot translates $\xi_k \pm T, \xi_k \pm 2T, \ldots$ in the intervals $D \pm T, D \pm 2T, \ldots$ with $D = [0, T)$ are identical. Hence, $\xi_{k+M} = \xi_k + T$, $k \in \mathbb{Z}$, for some M. Now let $p = \sum_k c_k b_k$ be a T-periodic spline in S^n_ξ. Then, $p(x) = p(x - T)$ implies

$$\sum_k c_k b_k(x) = \sum_k c_k b_{k+M}(x), \quad x \in \mathbb{R},$$

since $b_k(x - T) = b_{k+M}(x)$ by the periodicity of the knots. Shifting the index k by M, the sum on the right side becomes $\sum_k c_{k-M} b_k(x)$. Hence, $c_{k-M} = c_k$ by the uniqueness of the B-spline representation.

☐ **Example:**
We consider 2-periodic splines with simple knots at the integers ($\xi = \mathbb{Z}$). Since

$$\dim S^n_{(0,1),2} = 2,$$

and because of the periodicity condition $c_{k+2} = c_k$, a basis is formed by the constant spline ($c_k = 1$) and the oscillating spline

$$p = \sum_k (-1)^k b^n_{k,\xi}.$$

The latter has an interesting property. By the differentiation formula for standard uniform B-splines, $(b^n_{k,\xi})' = (b^{n-1}_{k,\xi} - b^{n-1}_{k+1,\xi})$, and, therefore,

$$p' = \sum_k \left((-1)^k - (-1)^{k-1}\right) b^{n-1}_{k,\xi}.$$

Hence, up to a factor, all derivatives of p have the same coefficient sequence:

$$p^{(m)} = 2^m \sum_k (-1)^k b^{n-m}_{k,\xi}.$$

4.6. Periodic Splines

Moreover, p satisfies the functional equation

$$p(x+1) = -p(x), \quad x \in \mathbb{R}.$$

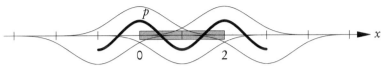

The figure shows p for degree $n = 3$ as well as the relevant B-splines $b_{k,\xi}^3$ for the periodicity interval $[0,2)$ multiplied by their coefficients. □

Chapter 5

Approximation

B-splines facilitate the analysis and implementation of approximation methods. With by now standard techniques, one can easily show that the smoothness constraints do not significantly affect the accuracy of local polynomial approximations. As a result, numerous approximation methods have been developed to suit any type of application.

As a very simple but instructive example of B-spline approximations, we discuss Schoenberg's scheme in Section 5.1. It illustrates the concept of quasi-interpolation described in Section 5.2. With the aid of quasi-interpolants we prove in Section 5.3 that splines approximate smooth functions with optimal order. Another application is the proof of de Boor's famous stability theorem in Section 5.4. The final two sections are devoted to spline interpolation. After discussing the well-posedness of general interpolation problems in Section 5.5, we consider the two principal smoothing techniques in Section 5.6.

5.1 ▪ Schoenberg's Scheme

Schoenberg's scheme is a generalization of the Bernstein operator. Just by sampling a function f at the appropriate points, an approximation is obtained which not only has linear precision but also preserves the shape of the graph of f.

Schoenberg's Scheme

Schoenberg's scheme uses function values at the knot averages $\xi_k^n = (\xi_{k+1} + \cdots + \xi_{k+n})/n$ as coefficients of a spline approximation to a smooth function f:

$$f \mapsto Qf = \sum_{k=0}^{m-1} f(\xi_k^n) b_k \in S_\xi^n$$

with $\xi : \xi_0, \dots, \xi_{m+n}$.

The method is second order accurate; i.e., for $x \in [\xi_\ell, \xi_{\ell+1}] \subseteq D_\xi^n = [\xi_n, \xi_m]$,

$$|f(x) - Qf(x)| \leq \frac{1}{2} \|f''\|_{\infty, D_x} h(x)^2,$$

where $D_x = [\xi_{\ell-n}^n, \xi_\ell^n]$, $\|f''\|_{\infty, D_x}$ denotes the maximum norm of f'' on D_x, and $h(x) = \max(\xi_\ell^n - x, x - \xi_{\ell-n}^n)$.

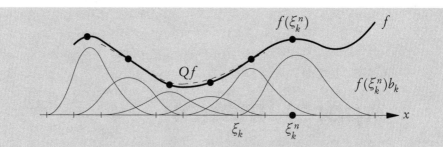

The Schoenberg operator preserves positivity, monotonicity, and convexity. This means that

$$f^{(k)} \geq 0 \implies (Qf)^{(k)} \geq 0$$

for $k \leq 2$ if both derivatives are continuous. For a uniform knot sequence, the sign of all derivatives up to order n is preserved.

We first observe that Schoenberg's scheme is exact for linear functions. Indeed, by Marsden's identity restricted to the parameter interval D_ξ^n

$$1 = \sum_k b_k(x), \quad x = \sum_k \xi_k^n b_k(x), \quad x \in D_\xi^n.$$

Therefore, $p = Qp$ for $p(x) = \alpha + \beta x$ in view of the linearity of Q.

With p the linear Taylor polynomial to f at x, it follows that

$$f(x) - (Qf)(x) = \underbrace{f(x) - p(x)}_{=0} + (Q(p-f))(x).$$

Hence, the absolute value of the error is

$$\leq \sum_k |(p-f)(\xi_k^n)| b_k(x) \leq \max_{\ell - n \leq k \leq \ell} |(p-f)(\xi_k^n)|$$

since the B-splines are positive and sum to one and at most $n+1$ B-splines are nonzero at x. Finally, by the formula for the Taylor remainder,

$$(p-f)(\xi_k^n) = \frac{1}{2} f''(y)(\xi_k^n - x)^2,$$

with y between x and ξ_k^n. Noting that $y \in D_x$ and $\xi_{\ell-n}^n \leq x \leq \xi_\ell^n$, the desired estimate follows.

It remains to show that Schoenberg's scheme is shape-preserving. To this end, we consider each derivative in turn. Clearly,

$$f \geq 0 \implies Qf = \sum_k f(\xi_k^n) b_k \geq 0$$

5.1. Schoenberg's Scheme

because of the positivity of the B-splines. For the first derivative, we obtain with the aid of the differentiation formula

$$(Qf)' = \sum_k \frac{f(\xi_k^n) - f(\xi_{k-1}^n)}{(\xi_{k+n} - \xi_k)/n} \, b_{k,\xi}^{n-1}.$$

The denominators of the coefficients equal $\xi_k^n - \xi_{k-1}^n$ by definition of the knot averages. Hence, by the mean value theorem,

$$(Qf)' = \sum_k f'(y_k) \, b_{k,\xi}^{n-1},$$

which is positive if f' is.

Applying the differentiation formula once more, $(Qf)'' = \sum_k c_k \, b_{k,\xi}^{n-2}$ with

$$c_k = \frac{n-1}{\xi_{k+n-1} - \xi_k} \left(\frac{f(\xi_k^n) - f(\xi_{k-1}^n)}{\xi_k^n - \xi_{k-1}^n} - \frac{f(\xi_{k-1}^n) - f(\xi_{k-2}^n)}{\xi_{k-1}^n - \xi_{k-2}^n} \right)$$

$$= \frac{(n-1)(\xi_k^n - \xi_{k-2}^n)}{\xi_{k+n-1} - \xi_k} \Delta(\xi_{k-2}^n, \xi_{k-1}^n, \xi_k^n) f,$$

where $\Delta(t_0, t_1, t_2)$ denotes the second divided difference at the points t_ν. Hence, by the generalized mean value theorem,

$$c_k = \varrho_k \frac{1}{2} f''(z_k)$$

with $\varrho_k > 0$. It follows that the sign of the second derivative is also preserved.

For higher derivatives, we obtain similar formulas only if the factors ϱ_k do not depend on k, which is the case for a uniform knot sequence ξ. Then,

$$(Qf)^{(m)} = h^{-m} \sum_k \nabla^m f(\xi_k^n) \, b_{k,\xi}^{n-m},$$

where h is the grid width and ∇ denotes the backward difference (with respect to the index k). By the mean value theorem for repeated differences, the coefficients $h^{-m} \nabla^m f(\xi_k^n)$ equal $f^{(m)}(z_k)$ for some points $z_k \in (\xi_{k-m}^n, \xi_k^n)$. As a consequence, the sign of all existing derivatives is preserved.

To achieve optimal accuracy, the placement of the knots is crucial. Intuitively, more knots are needed where the second derivative of the approximated function is large. This is illustrated in the following example.

□ **Example:**

We apply Schoenberg's scheme to the function $f(x) = \sqrt{x}$, $x \in [0,1]$, using quadratic splines. The left figure shows the error of the approximation Qf for uniform knots:

$$-h = \xi_0, \, \xi_1 = 0 = \xi_2, \, \xi_3 = h, \, \ldots, \, \xi_k = (k-2)h, \, \ldots, \, \xi_m = 1 = \xi_{m+1}, \, \xi_{m+2} = 1 + h,$$

where $h = 1/(m-2) \asymp 1/m$. As is to be expected, the error grows for $x \to 0$, where the derivatives of f become infinite. It can be shown that the maximum of the error merely decays like $O(1/\sqrt{m})$.

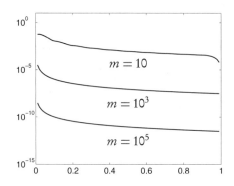
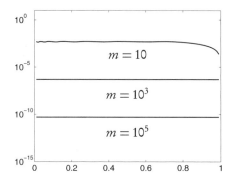

The unbalanced distribution of the error can be avoided by placing more knots near $x = 0$. With the choice

$$\xi_k = ((k-2)/(m-2))^4, \quad k = 3, \ldots, m-1,$$

of the interior knots, the error is approximately constant over the interval D_ξ^2, as is shown in the right figure. We claim that

$$|f(x) - Qf(x)| \lesssim m^{-2}, \quad x \in D_\xi^2,$$

where the symbol $\lesssim$ indicates inequalities up to constant factors which do not depend on m. Hence, without increasing the dimension of S_ξ^2, we have gained a factor $O(m^{-3/2})$.

To verify this estimate, we consider two cases.

For $0 \le x < \xi_4$, only B-splines b_k with $k < 4$ are relevant. Since the corresponding knot averages ξ_k^2 as well as x are less than $\xi_5 = (3/(m-2))^4$,

$$\left| \sqrt{x} - \sum_{k=0}^{3} \sqrt{\xi_k^2}\, b_k(x) \right| \lesssim \sqrt{m^{-4}},$$

as desired.

For $\xi_\ell \le x < \xi_{\ell+1}$ with $\ell \ge 4$, we can apply the error estimate for Schoenberg's scheme since the second derivative of the square root is bounded by

$$\frac{1}{4}\left(\xi_{\ell-2}^2\right)^{-3/2} \le \frac{1}{4}\left(\xi_{\ell-1}\right)^{-3/2} = \frac{1}{4}\left(\frac{\ell-3}{m-2}\right)^{4(-3/2)} \lesssim (\ell/m)^{-6}$$

on the interval $D_x = [\xi_{\ell-2}^2, \xi_\ell^2] \subset [\xi_{\ell-1}, \xi_{\ell+2}]$. Since

$$h(x) \le \xi_\ell^2 - \xi_{\ell-2}^2 \le \xi_{\ell+2} - \xi_{\ell-1} = (\ell^4 - (\ell-3)^4)/(m-2)^4 \lesssim \ell^3/m^4,$$

5.2. Quasi-Interpolation

the absolute value of the error is

$$\leq \frac{1}{2}\|f''\|_{\infty,D_x} h(x)^2 \lesssim \frac{1}{2}(\ell/m)^{-6}(\ell^3/m^4)^2,$$

proving the estimate also in this case. □

The second example illustrates the shape preservation of Schoenberg's scheme. Clearly, the exponential function, with identical positive derivatives of all orders, provides an ideal test case.

□ **Example:**
We apply Schoenberg's scheme with uniform knots with grid width h to $f(x) = \exp(x)$ for $x \in [0,1]$. The left figure shows the derivatives

$$(Qf)^{(j)}, \quad j \leq n,$$

for degree $n = 3$, confirming the shape preservation of the method. Since $f^{(j)}(x) = \exp(x) \geq 0$, all derivatives of Schoenberg's approximation Qf are monotone increasing. Except for the third derivative, which is piecewise constant, they can hardly be distinguished from the function itself.

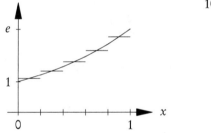
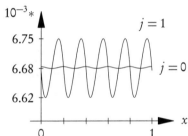

As shown in the right figure, the absolute value of the relative error

$$|(Qf)^{(j)}(x) - \exp(x)|/\exp(x)$$

is less than 1% for $j = 0$ and $j = 1$, although only five grid intervals are used ($h = 1/5$). □

5.2 ▪ Quasi-Interpolation

For an orthonormal basis $b_0, b_1, \ldots$, the orthogonal projection

$$f \sim \sum_k c_k b_k, \quad c_k = \langle f, b_k \rangle,$$

provides an optimal approximation. Unfortunately, B-splines are not orthogonal unless one modifies the standard integral inner product as was suggested by Reif. Nevertheless, it is possible to construct approximation processes which are almost as efficient. The coefficients are computed via local linear functionals rather than by forming scalar products.

Quasi-Interpolant
A linear spline approximation scheme

$$f \mapsto Qf = \sum_k (Q_k f)\, b_k \in S_\xi^n$$

for continuous functions is called a quasi-interpolant of maximal order if the following conditions are satisfied.

(i) Q_k are locally bounded linear functionals, i.e.,

$$|Q_k f| \le \|Q\| \|f\|_{\infty,[\xi_k,\xi_{k+n+1}]},$$

where $\|f\|_{\infty,U} = \max_{x \in U} |f(x)|$.

(ii) Q reproduces polynomials p of degree $\le n$, i.e., $Qp = p$ on the parameter interval D_ξ^n of S_ξ^n. Equivalently, for $y \in \mathbb{R}$,

$$Q_k p = \psi_k(y), \quad p(x) = (x-y)^n$$

with $\psi_k(y) = (\xi_{k+1} - y) \cdots (\xi_{k+n} - y)$.

The characterization of the reproduction of polynomials is an immediate consequence of Marsden's identity, which asserts that

$$(x-y)^n = \sum_k \psi_k(y) b_k(x).$$

Since the representation of any polynomial of degree $\le n$ as linear combination of B-splines is unique,

$$p = Qp = \sum_k (Q_k p)\, b_k$$

with $p(x) = (x-y)^n$ implies that $Q_k p = \psi_k(y)$. Moreover, since the polynomials $(\cdot - y)^n$, $y \in \mathbb{R}$, span $\mathbb{P}^n$, this condition is not only necessary but also sufficient for polynomial reproduction.

We have introduced quasi-interpolation in the simplest setting. Referring to the work by de Boor, Lyche and Schumaker, and Fix, the concept is much more general. Obviously, different norms can be used. Moreover, we need not insist on maximal order, i.e., require merely that polynomials of degree $\le m$ with $m < n$ are reproduced. This is the case for Schoenberg's scheme, which has linear precision ($m = 1$) and served as an introductory example. While not particularly difficult, every generalization makes the analysis slightly more technical. Hence, we have confined ourselves to the case which is perhaps of most practical importance.

☐ **Example:**
We construct a quasi-interpolant for splines with the uniform knot sequence $\xi = h\mathbb{Z}$. A natural choice for the functionals is

$$Q_k f = \sum_{v=0}^{n} w_v f((k+1/2+v)h),$$

i.e., a weighted sum of function values at the midpoints of the knot intervals. This scheme is particularly efficient since neighboring functionals share common data.

The coefficients w_ν are determined from the condition for the reproduction of polynomials:

$$\sum_{\nu=0}^{n} w_\nu ((k+1/2+\nu)h - y)^n = \prod_{\alpha=1}^{n} ((k+\alpha)h - y).$$

Testing this polynomial equation for $y = (k+1/2+\mu)h$ leads to the linear system

$$\sum_{\nu=0}^{n} w_\nu (\nu-\mu)^n = \prod_{\alpha=1}^{n} (\alpha - 1/2 - \mu), \quad \mu = 0,\ldots,n,$$

after cancelling the factor h^n on both sides. It follows that the coefficients, defining the functionals Q_k, depend neither on k nor on h. Hence, the functionals are uniformly bounded,

$$\|Q\| = \sum_\nu |w_\nu|,$$

which is the other requirement for a quasi-interpolant.

	w_0	w_1	w_2	w_3	w_4
$n=1$	$\dfrac{1}{2}$	$\dfrac{1}{2}$			
$n=2$	$-\dfrac{1}{8}$	$\dfrac{5}{4}$	$-\dfrac{1}{8}$		
$n=3$	$-\dfrac{7}{48}$	$\dfrac{31}{48}$	$\dfrac{31}{48}$	$-\dfrac{7}{48}$	
$n=4$	$\dfrac{47}{1152}$	$-\dfrac{107}{288}$	$\dfrac{319}{192}$	$-\dfrac{107}{288}$	$\dfrac{47}{1152}$

The table shows the coefficients up to degree 4. We observe that this quasi-interpolant is not a projection. For example,

$$Q_k b_{k+n} = w_n b_{k+n}(kh + h/2 + nh) \neq 0,$$

while, for a projection, $Q b_\ell = b_\ell$, which implies $Q_k b_\ell = \delta_{k,\ell}$. $\qquad\square$

With many different types of quasi-interpolants at our disposal, the question of optimality arises. While there is no unique answer, two important additional features are reproduction of B-splines (not just polynomials) and uniform boundedness with respect to knot refinement.

Standard Projector

A quasi-interpolant

$$f \mapsto Qf = \sum_k (Q_k f)\, b_k \in S_\xi^n,$$

for which each linear functional Q_k, $k \sim \xi$, depends only on values of f in a single knot interval in the parameter interval D_ξ^n of S_ξ^n, is a projector, i.e.,

$$Qp = p \quad \forall p \in S_\xi^n.$$

Such quasi-interpolants are called standard projectors if the norms of the linear functionals can be bounded by a constant $\|Q\|$ which depends only on the degree n. Projectors of this type exist if all B-splines have a largest knot interval of their support in D_ξ^n.

According to the condition for a projector, assume that $Q_k p$ depends only on $p_{|D_\ell}$, $D_\ell = [\xi_\ell, \xi_{\ell+1}]$, with $\ell = \ell(k)$. In order to verify that Q reproduces any spline $p = \sum_j c_j\, b_j \in S_\xi^n$, we must show that

$$Q_k p = c_k$$

for all $k \sim \xi$. To this end, let q be the polynomial which agrees with p on D_ℓ:

$$p(x) = \sum_{j=\ell-n}^{\ell} c_j\, b_j(x) = q(x), \quad x \in D_\ell.$$

Hence,

$$Q_k p = Q_k q = c_k$$

since Q reproduces polynomials and the coefficients in the representation of a polynomial q are unique.

We now construct a particular projector with $\|Q\|$ independent of ξ by choosing

$$Q_k f = \sum_{v=0}^{n} w_v\, f(\xi_\ell + vh), \quad h = (\xi_{\ell+1} - \xi_\ell)/n,$$

where $[\xi_\ell, \xi_{\ell+1}] \subseteq D_\xi^n$ is a largest knot interval in the support of b_k. The coefficients w_v are determined by the condition for exactness of Q on $\mathbb{P}^n$:

$$\sum_v w_v(\xi_\ell + vh - y)^n = (\xi_{k+1} - y)\cdots(\xi_{k+n} - y) \quad \forall y \in \mathbb{R}.$$

In view of the uniqueness of polynomial interpolation, it is sufficient to test this equation between polynomials of degree $\leq n$ for $y = \xi_\ell + \mu h$, $\mu = 0, \ldots, n$. After division by h^n, this leads to the linear system

$$\sum_v w_v(v - \mu)^n = ((\xi_{k+1} - \xi_\ell)/h - \mu)\cdots((\xi_{k+n} - \xi_\ell)/h - \mu), \quad \mu = 0, \ldots, n.$$

Since $[\xi_\ell, \xi_{\ell+1}]$ is a largest interval of the support of b_k, the length of the interval $[\xi_k, \xi_{k+n+1}]$ can be compared with h:

$$\xi_{k+n+1} - \xi_k \leq (n+1)(\xi_{\ell+1} - \xi_\ell) = (n+1)n\, h.$$

Consequently, each factor on the right side of the linear system is $\leq c(n)$. The same estimate holds, with a different constant $c(n)$, for its solution w. Therefore,

$$|Q_k f| \leq c(n) \max_{v} |f(\xi_\ell + vh)|,$$

establishing the knot independent bound $\|Q\|$.

For a projector Q we have $Q b_k = b_k$ for any B-spline b_k. This implies that

$$Q_k b_j = \delta_{k,j};$$

i.e., the functionals Q_k are dual to the B-splines. Such dual functionals were first constructed by de Boor. Numerous variants exist and they play a crucial role in the analysis of splines.

□ **Example:**

We construct a standard projector for quadratic splines. Choosing $[\xi_\ell, \xi_{\ell+1}]$ as the middle interval of the support of b_k,

$$Q_k f = w_0 f(\xi_{k+1}) + w_1 f(\eta_k) + w_2 f(\xi_{k+2}), \quad \eta_k = (\xi_{k+1} + \xi_{k+2})/2.$$

Then, as for the general case, Marsden's identity yields the linear system

$$
\begin{array}{ccccccc}
0 & + & w_1 & + & 4w_2 & = & 0, \\
w_0 & + & 0 & + & w_2 & = & -1, \\
4w_0 & + & w_1 & + & 0 & = & 0.
\end{array}
$$

For example, the middle entry (-1) on the right side equals

$$((\xi_{k+1} - \xi_\ell)/h - 1)((\xi_{k+2} - \xi_\ell)/h - 1)$$

with $\ell = k+1$ and $h = (\xi_{k+2} - \xi_{k+1})/2$. The solution

$$w_0 = w_2 = -\frac{1}{2}, \quad w_1 = 2$$

is independent of the knot sequence. In particular,

$$|Q_k f| \leq \left(\frac{1}{2} + 2 + \frac{1}{2} \right) \max_{\xi_{k+1} \leq x \leq \xi_{k+2}} |f(x)|,$$

so that $\|Q\| = 3$. □

5.3 ▪ Accuracy of Quasi-Interpolation

Taylor polynomials of degree $\leq n$ approximate smooth functions locally with order $n+1$. This accuracy is maintained by splines regardless of their smoothness and the placement of the knots.

Accuracy of Quasi-Interpolation
The error of a quasi-interpolant

$$f \mapsto Qf = \sum_k (Q_k f) b_k \in S_\xi^n$$

of maximal order satisfies

$$|f(x) - (Qf)(x)| \leq \frac{\|Q\|}{(n+1)!} \|f^{(n+1)}\|_{\infty, D_x} h(x)^{n+1}, \quad x \in D_\xi^n,$$

where D_x is the union of the supports of all B-splines b_k, $k \sim x$, which are relevant for x, and $h(x) = \max_{y \in D_x} |y - x|$.

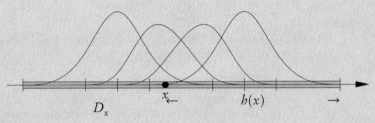

If the local mesh ratio is bounded, i.e., if the quotients of the lengths of adjacent knot intervals are $\leq r_\xi$, then the error of the derivatives on the knot intervals $(\xi_\ell, \xi_{\ell+1})$ can be estimated by

$$|f^{(j)}(x) - (Qf)^{(j)}(x)| \leq c(n, r_\xi) \|Q\| \|f^{(n+1)}\|_{\infty, D_x} h(x)^{n+1-j}$$

for all $j \leq n$.

Choosing a standard projector for Q shows, in particular, that splines approximate smooth functions with optimal order; the norm $\|Q\|$ does depend only on n in this case.

The by now standard proof is elegant and simple—an almost immediate consequence of the defining properties of quasi-interpolants.

With p the Taylor polynomial of degree $\leq n$ to f at x, it follows that

$$f(x) - (Qf)(x) = \underbrace{f(x) - p(x)}_{=0} + (Q(p-f))(x)$$

since Q reproduces polynomials. Hence, the absolute value of the error is

$$\leq \sum_k |Q_k(p-f)| b_k(x) \leq \max_{k \sim x} |c_k|, \quad c_k = Q_k(p-f),$$

since the B-splines sum to one, and the maximum needs to be taken only over the relevant indices, i.e., k with $b_k(x) > 0$. Using the boundedness of the quasi-interpolant,

$$|c_k| \leq \|Q\| \|p - f\|_{\infty, [\xi_k, \xi_{k+n+1}]}.$$

Moreover, by the formula for the Taylor remainder,

$$|(p-f)(y)| = \frac{1}{(n+1)!} |f^{(n+1)}(z)| |y - x|^{n+1}, \quad \xi_k \leq y \leq \xi_{k+n+1},$$

with z between x and y. Observing that $y, z \in D_x$, which implies, in particular, that $|y - x| \le h(x)$,

$$|c_k| \le \frac{\|Q\|}{(n+1)!} \|f^{(n+1)}\|_{\infty, D_x} h(x)^{n+1}, \quad k \sim x,$$

and the desired estimate follows.

The error of the derivatives is estimated in an analogous fashion. Considering the first derivative, it suffices to bound the absolute value of

$$e = (Q(p-f))' = \left(\sum_{k=\ell-n}^{\ell} c_k b_k\right)'$$

on the knot interval $(\xi_\ell, \xi_{\ell+1})$ containing x. Applying the differentiation formula,

$$e = \sum_{k=\ell-n+1}^{\ell} \alpha_{k,\xi}^n (c_k - c_{k-1}) b_{k,\xi}^{n-1}, \quad \alpha_{k,\xi}^n = \frac{n}{\xi_{k+n} - \xi_k}.$$

Since, for the relevant range of indices,

$$\xi_k \le \xi_\ell, \quad \xi_{k+n} \ge \xi_{\ell+1},$$

differentiation increases the size of the maximal absolute value of the B-spline coefficients at most by a factor $2n/|\xi_{\ell+1} - \xi_\ell|$. Therefore,

$$|e(x)| \le \frac{2n}{\xi_{\ell+1} - \xi_\ell} \max_{k \sim x} |c_k|.$$

Using the previously derived estimate for the coefficients c_k yields the desired bound for the error if $h(x) \le c(\xi_{\ell+1} - \xi_\ell)$. This follows because the size of neighboring knot intervals can be compared in terms of the local mesh ratio. Since D_x is contained in the union of n intervals to the left and to the right of $[\xi_\ell, \xi_{\ell+1}]$,

$$\frac{h(x)}{\xi_{\ell+1} - \xi_\ell} \le (1 + r_\xi + r_\xi^2 + \cdots + r_\xi^n) \le c(n, r_\xi),$$

as required.

For the jth derivative, the B-spline coefficients of $(Q(p-f))^{(j)}$ are obtained by forming repeated weighted differences of the coefficients c_k according to the differentiation formula. Similarly, as explained above, each difference operation increases the maximal absolute value of the coefficients at most by a factor $2(n+1-j)/(\xi_{\ell+1} - \xi_\ell)$. Hence, the error of the jth derivative is at most by a factor $c(n, r_\xi)/h(x)^j$ larger than the bound for $|f(x) - (Qf)(x)|$.

□ **Example:**

We approximate the periodic function

$$f(x) = \sin\left(\frac{1}{1.1 + \sin(x)}\right)$$

on $D = [0, 2\pi]$ with a cubic quasi-interpolant, based on evaluation at the midpoints of the knot intervals.

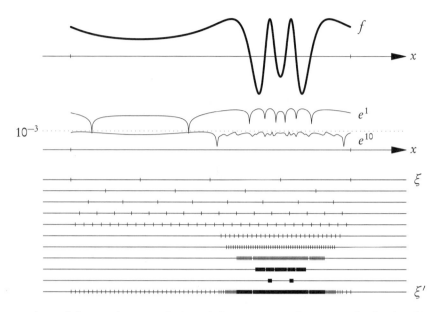

In view of the varying complexity of f, we use an adaptive method. To achieve an error less that $\varepsilon = 10^{-3}$, we start with the uniform knot sequence

$$\xi : \ldots, 0, \pi/2, \pi, \ldots$$

and place additional knots at the midpoints of the knot intervals where the maximal error is $\geq \varepsilon$. As is illustrated in the figure, nine steps suffice to reduce the error e below the prescribed bound. The final knot sequence ξ', shown at the bottom of the figure, consists of 337 knots in D. □

5.4 ▪ Stability

The B-spline basis is well conditioned; i.e., small changes in the coefficients result in small changes of the corresponding spline function and vice versa. This fundamental stability result, due to de Boor, not only is important for numerical computations but also has several theoretical applications.

Stability
The size of a spline

$$p = \sum_k c_k b_k \in S_\xi^n$$

is comparable with the size of its coefficients:

$$c(n) \sup_k |c_k| \leq \sup_{x \in D_\xi^n} |p(x)| \leq \sup_k |c_k|.$$

The constant depends on the degree n. It does not depend on the knot sequence ξ if the parameter interval D_ξ^n of S_ξ^n contains for each B-spline a largest knot interval in its support.

Denote by b_k, $k \sim x$, the B-splines which are nonzero at x. Since the B-splines are positive and sum to one,

$$|p(x)| \le \sum_{k \sim x} |c_k| \, b_k(x) \le \left(\max_{k \sim x} |c_k| \right) \underbrace{\sum_{k \sim x} b_k^n(x)}_{=1},$$

and we obtain the upper bound.

To prove the lower bound, we use a standard projector

$$f \mapsto Qf = \sum_k (Q_k f) \, b_k$$

with

$$|Q_k f| \le \|Q\| \max_{\xi_\ell \le x \le \xi_{\ell+1}} |f(x)|, \quad [\xi_\ell, \xi_{\ell+1}] \subseteq D_\xi^n,$$

where $\ell = \ell(k)$ and $\|Q\|$ depends only on the degree n. As we remarked in Section 5.2,

$$Q_k b_j = \delta_{k,j}.$$

Therefore,

$$c_k = Q_k p,$$

and the uniform bound for the functionals Q_k implies the lower bound with $c(n) = 1/\|Q\|$.

The hypothesis for the stability inequalities is not a severe restriction since we are free to choose the exterior knots. Modifying knots $\xi_k \notin D_\xi^n$ does not change the spline space S_ξ^n. Nevertheless, a restriction on the exterior knots is necessary. To explain this, we consider the spline space S_ξ^2 with

$$\xi: \quad \xi_0 = \xi_1 = -h, \; \xi_2 = 0, \; \xi_3 = 1, \; \xi_4 = \xi_5 = 2.$$

Then, for $x \in D_\xi^2 = [0,1]$,

$$b_{0,\xi}^2(x) = \frac{(1-x)^2}{1+h}.$$

Hence, for $p = c_0 b_{0,\xi}^2$ with $c_0 = 1$,

$$\|p\|_{\infty,[0,1]} = 1/(1+h) \to 0, \quad h \to \infty,$$

while $\max_k |c_k| = 1$ for any h, i.e., the lower bound of the stability result is violated.

□ **Example:**

As an application of the stability inequalities, we estimate the size of the derivative of a spline

$$p = \sum_{k=0}^{m-1} c_k \, b_k \in S_\xi^n$$

with $\xi : \xi_0, \dots, \xi_{m+n}$ and knot multiplicities $< n$. By the differentiation formula,

$$p' = \sum_{k=1}^{m-1} c_k' \, b_{k,\xi}^{n-1}, \quad c_k' = \frac{n(c_k - c_{k-1})}{\xi_{k+n} - \xi_k},$$

on the parameter interval $D_\xi^n = [\xi_n, \xi_m]$. Taking absolute values, we obtain

$$\max_{x \in D_\xi^n} |p'(x)| \le \frac{2n}{\min\limits_{0 < k < m} (\xi_{k+n} - \xi_k)} \max_{0 \le k < m} |c_k|.$$

Hence, by the stability inequality,

$$\|p'\|_{\infty, D_\xi^n} \le \frac{c(n)}{h_-} \|p\|_{\infty, D_\xi^n},$$

where h_- is the minimal length of a knot interval in D_ξ^n. $\qquad\square$

5.5 ▪ Interpolation

Interpolation is a simple and popular approximation method. Moreover, for many applications, it is essential to match given data exactly. However, unlike for polynomials, the local support of the B-splines implies that there must be some restrictions on the placement of the interpolation points. It is surprising, though, that the characterization of admissible point sets has a very simple form.

Before formulating the main theorem, some notation for specifying general Hermite interpolation conditions is convenient. We say that p interpolates f at a nondecreasing sequence of points $t : \ldots \le t_0 \le t_1 \le \ldots$ if

$$p^{(\mu_j)}(t_j) = f^{(\mu_j)}(t_j),$$

where μ_j is the largest integer i with $t_{j-i} = t_j$. In other words, at a point t_j with multiplicity μ, all derivatives up to order $\mu - 1$ are interpolated. For points t_j with multiplicity 1 ($\mu_j = 0$) only the values match: $p(t_j) = f(t_j)$. Moreover, we say that p vanishes at t if all interpolation data are zero:

$$p^{(\mu_j)}(t_j) = 0 \quad \forall j.$$

For example, p vanishes at $t : t_0 = 0, 0, 1 = t_2$ means that $p(0) = p'(0) = p(1) = 0$. In this case, $\mu_0 = 0 = \mu_2$ and $\mu_1 = 1$.

Using multiple points to describe Hermite conditions combines well with the concept of knot sequences. However, some restrictions are necessary for interpolation with splines $p \in S_\xi^n$. Clearly, the multiplicities of the interpolation points should be $\le n + 1$; a derivative of order $n + 1$ cannot be interpolated with polynomials of degree $\le n$. Moreover, referring to the interpolation conditions above, it is reasonable to assume that the derivatives $p^{(\mu_j)}$ are continuous at t_j. This means that

$$\xi_k = t_j \implies \#\xi_k + \#t_j \le n + 1$$

in view of the smoothness conditions for splines; p is $n - \#\xi_k$ times continuously differentiable at ξ_k. A point sequence t which meets both of these requirements is called admissible.

With the above notations the characterization of well-posed spline interpolation problems can be stated in a rather elegant form.

5.5. Interpolation

Schoenberg–Whitney Conditions

For a spline space S_ξ^n with finite knot sequence $\xi_0, \ldots, \xi_{m+n}$ and an admissible non-decreasing sequence $t: t_0 \leq \cdots \leq t_{m-1}$ of interpolation points in the parameter interval D_ξ^n there exists a unique interpolating spline $p = \sum_{k=0}^{m-1} c_k b_k \in S_\xi^n$ for arbitrary data iff

$$\xi_k < t_k < \xi_{k+n+1}$$

for $k = 0, \ldots, m-1$.

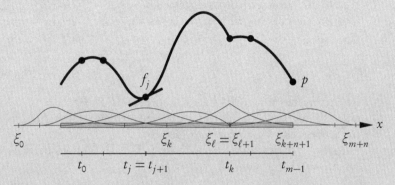

By linearity, the B-spline coefficients are determined by the linear system

$$Ac = f, \quad a_{j,k} = b_k^{(\mu_j)}(t_j), f_j = f^{(\mu_j)}(t_j).$$

Since on any knot-interval in D_ξ^n only $n+1$ B-splines are nonzero, the interpolation matrix A is banded; each row can have at most $n+1$ nonzero entries.

Our proof essentially follows de Boor, who derived the total positivity of the interpolation matrix, a slightly more general result.

In order to show that the Schoenberg–Whitney conditions are necessary, we assume, e.g., that $\xi_{\ell+n+1} \leq t_\ell$ for some ℓ. Then,

$$b_k^{(\mu_j)}(t_j) = 0, \quad k \leq \ell, j \geq \ell,$$

since t_j lies to the right of (ξ_k, ξ_{k+n+1}). This implies that the $m \times m$ interpolation matrix A contains a zero submatrix (the lower left block) of dimension $(m-\ell) \times (\ell+1)$, i.e., for columns 1 up to $\ell+1$ only the first ℓ entries can be nonzero. Hence, these columns are linearly dependent, so that A is singular. An analogous argument is possible if $t_\ell \leq \xi_\ell$ for some ℓ.

To establish the sufficiency we use induction on the degree n as well as on the dimension $m > n$ of the spline space.

We start with degree $n = 1$ and $m = 2$ points, recalling that piecewise constant functions ($n = 0$) are not considered as splines. Because of the continuity requirement for splines all four knots $\xi_0, \ldots, \xi_3$ are simple and admissible points $t_k \in D_\xi^1 = [\xi_1, \xi_2]$ satisfy

$$\xi_1 \leq t_0 < t_1 \leq \xi_2 \quad \vee \quad \xi_1 < t_0 = t_1 < \xi_2$$

since the common multiplicity of the sequences ξ and t must be less than or equal to $n+1 = 2$. In both cases it is obvious that unique interpolation at t with linear functions is possible on D_ξ^1.

We now turn to the induction step, considering degree n and $m > n$ points. In order to prove that the Schoenberg–Whitney conditions

$$\xi_k < t_k < \xi_{k+n+1}$$

imply the well-posedness of the interpolation problem, we can assume that this assertion has already been established for

$$n' < n \quad \lor \quad n' = n \land m' < m .$$

We can also assume that the exterior knots $\xi_0, \dots, \xi_n$ and $\xi_m, \dots, \xi_{m+n}$ are simple since their position has no effect on the spline space S_ξ^n. Now, by linearity, it suffices to show that a spline

$$p = \sum_{k=0}^{m-1} c_k b_k$$

which vanishes at t must be zero. Equivalently, $Ac = 0 \implies c = 0$ for the $m \times m$ interpolation matrix A. To this end, we consider three cases, adding additional inequalities to the Schoenberg–Whitney conditions.

(i) $t_{\ell-1} \le \xi_\ell$ for some interior knot $\xi_\ell > \xi_{\ell-1}$: We consider the restrictions of p to the subintervals $[\xi_n, \xi_\ell]$ and $[\xi_\ell, \xi_m]$ of D_ξ^n,

$$q = \sum_{k=0}^{\ell-1} c_k b_k, \quad r = \sum_{k=\ell+\mu-1-n}^{m-1} c_k b_k ,$$

where μ denotes the multiplicity of ξ_ℓ, i.e.,

$$\xi_{\ell-1} < \xi_\ell = \cdots = \xi_{\ell+\mu-1} < \xi_{\ell+\mu} .$$

Note that the B-splines b_k with $k < \ell+\mu-1-n$ vanish on the right subinterval $[\xi_\ell, \xi_m]$, explaining the range of summation in the expression for r.

The two splines q and r belong to the spline spaces S_η^n, S_ζ^n with

$$\eta : \xi_0, \dots, \xi_{\ell+n}, \quad \zeta : \xi_{\ell+\mu-1-n}, \dots, \xi_{m+n} .$$

Since $t_{\ell-1} \le \xi_\ell$, the restriction q of p to the left subinterval $[\xi_n, \xi_\ell]$ vanishes at

$$u : t_0, \dots, t_{\ell-1} .$$

Clearly, u is admissible for S_η^n and as subsequence of t satisfies the Schoenberg–Whitney conditions. Since $\dim S_\eta^n = \ell < m$, this implies $q = 0$ by induction hypothesis, i.e., $c_0 = \cdots = c_{\ell-1} = 0$.

Similarly, r vanishes at the point sequence

$$v : t_\ell, \dots, t_{m-1} ,$$

contained in the right interval $[\xi_\ell, \xi_m]$. In this case, the induction hypothesis is not directly applicable since the number of points, which equals $m-\ell$, is less than the dimension $m-\ell+n+1-\mu$ of S_ζ^n. However, since the coefficients $c_{\ell+\mu-1-n}, \dots, c_{\ell-1}$ of the B-splines which are nonzero at ξ_ℓ have already shown to be zero, we can augment ξ_ℓ with multiplicity $n+1-\mu$ to v as an additional interpolation point where r vanishes. This leads to the extended sequence

$$\tilde{v} : \xi_\ell, \dots, \xi_\ell, t_\ell, \dots, t_{m-1}$$

of correct length, equal to $\dim S_\zeta^n$. The admissibility of v is preserved, as is the validity of the Schoenberg–Whitney conditions; each of the B-splines $b_{\ell+\mu-1-n},\dots,b_{\ell-1}$ contains $\tilde{v}_{\ell+\mu-1-n}=\cdots=\tilde{v}_{\ell-1}=\xi_\ell$ in the interior of its support. Now, with r vanishing at a suitable point sequence $\tilde{v}$, $r=0$ by induction hypothesis, proving that also the remaining B-spline coefficients $c_\ell,\dots,c_{m-1}$ must be zero.

(ii) $\xi_\ell \leq t_{\ell-n}$ for some interior knot $\xi_\ell < \xi_{\ell+1}$: We conclude that $c_k = 0$ with completely analogous arguments as in the case (i); by symmetry, the role of left and right is reversed.

(iii) $t_{\ell-n} < \xi_\ell < t_{\ell-1}$ for all interior knots ξ_ℓ: This case is relevant in particular when there are no interior knots, i.e., if $m = n+1$ and S_ζ^n consists of polynomials of degree $\leq n$.

Rewriting the inequalities for ξ_ℓ in the form

$$\xi_\ell < t_{\ell-1} < \xi_{\ell+n-1}$$

implies that interior knots can at most have multiplicity $n-1$. Hence, p is continuously differentiable on D_ξ^n with

$$p' = \sum_{k=1}^{m-1} d_k b_{k,\xi'}^{n-1} \in S_{\xi'}^{n-1}, \quad \xi' : \xi_1,\dots,\xi_{m+n-1}.$$

With p vanishing at $t_0,\dots,t_{m-1}$, by Rolle's theorem, p' has at least $m-1$ zeros

$$t' : t_1' \leq \cdots \leq t_{m-1}',$$

counting multiplicities. More precisely, if s is a multiple zero of p, differentiation reduces the multiplicity by 1, and $p(s_1) = 0 = p(s_2)$ with $s_1 < s_2$ implies that $p'(s) = 0$ for some $s \in (s_1, s_2)$. Consequently, for appropriately chosen zeros t_ℓ',

$$t_\ell' = t_{\ell-1} = t_\ell \quad \vee \quad t_\ell' \in (t_{\ell-1}, t_\ell).$$

This shows that t' is admissible for $S_{\xi'}^{n-1}$; either the multiplicity is decreased or new points with multiplicity 1 are introduced. If t' also satisfies the Schoenberg–Whitney conditions, i.e., if

$$\xi_k < t_k' < \xi_{k+n}, \quad k = 1,\dots,m-1,$$

then the induction hypothesis is applicable. Vanishing at $m-1 = \dim S_{\xi'}^{n-1}$ points t' (counting multiplicities), the derivative p' is zero. This implies that p is zero, too, thereby completing the induction step also for case (iii).

The Schoenberg–Whitney conditions for t' are easily checked. The inequalities, characterizing the case under consideration, and $t_k' \in [t_{k-1}, t_k]$ imply

$$\xi_k < t_{k-1} \leq t_k', \quad k = n+1,\dots,m-1,$$
$$t_k' \leq t_k < \xi_{k+n}, \quad k = 1,\dots,m-n-1,$$

where the restrictions on k ensure that ξ_k and ξ_{k+n} are interior knots. It remains to establish the lower bounds for $t_1',\dots,t_n'$ and the upper bounds for $t_{m-n}',\dots,t_{m-1}'$ which are still missing in the above inequalities. By symmetry the arguments are similar. Considering, e.g., the lower bounds, we recall that the exterior knots are assumed to be simple,

$$\xi_0 < \cdots < \xi_n < \xi_{n+1},$$

and that all interpolation points lie in the parameter interval D_ξ^n,

$$\xi_n \leq t_0 \leq \cdots \leq t_{m-1} \leq \xi_m.$$

In view of $t_{k-1} \leq t'_k$ this yields the lower bound $\xi_k < t'_k$ for $k < n$. We also conclude that $\xi_n \leq t_{n-1} \leq t'_n$. However, equality in both inequalities cannot hold since $t_0 = \cdots = t_{n-1} = \xi_n$ and $t_{n-1} = t'_n = t_n$ would imply $\#\xi_n + \#t_n = 1 + (n+1)$, exceeding the maximal admissible multiplicity $n+1$. This establishes the Schoenberg–Whitney conditions for all t'_k, $k = 1, \ldots, m-1$, thus ensuring the applicability of the induction hypothesis, as claimed before.

◻ **Example:**
Frequently, one interpolates with cubic splines at simple knots in the parameter interval $D_\xi^3 = [\xi_3, \xi_m]$ of the spline space S_ξ^3. However, since $\dim S_\xi^3 = m$ and there are only $m - 2$ interpolation points, two additional conditions are required. A possible choice is the so-called not-a-knot condition, requiring that the third derivative of the interpolating spline is continuous at the first and last interior knots.

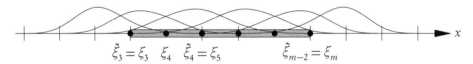

In effect, as is illustrated in the figure, the interpolation scheme uses B-splines $\tilde{b}_k$ with respect to the reduced knot sequence $\tilde{\xi}$ obtained from ξ by deleting the knots ξ_4 and ξ_{m-1}. Hence, the not-a-knot condition is incorporated in the spline space. With this formulation, the well-posedness of the linear system is an immediate consequence of the Schoenberg–Whitney conditions

$$\tilde{\xi}_k < x_k = \xi_{k+3} < \tilde{\xi}_{k+4}, \quad k = 0, \ldots, m-3,$$

which are obviously satisfied. ◻

◻ **Example:**
For uniform knots ξ_k with grid width h, a natural choice of interpolation points are the midpoints of the supports of the B-splines $b_k = b^n((\cdot - \xi_k)/h)$:

$$x_i = \xi_i + (n+1)h/2.$$

For odd (even) degree n, these points coincide with the knots (midpoints of the knot intervals). The corresponding nonzero entries

$$a_{i,k} = b^n((x_i - \xi_k)/h) = b^n(i - k + (n+1)/2)$$

of the interpolation matrix A are listed in the table below.

n	$a_{k,k}$	$a_{k\pm 1,k}$	$a_{k\pm 2,k}$
2	3/4	1/8	
3	2/3	1/6	
4	115/192	19/96	1/384
5	11/20	13/60	1/120

5.5. Interpolation

For splines on the real line, A is a Toeplitz matrix, i.e., $a_{i,k}$ depends only on the difference $i - k$. Similarly, for periodic splines with period $T = Mh$, $a_{i,k} = \alpha_{i-k \bmod M}$.

For a spline space S_ξ^n with parameter interval $D_\xi^n = [\xi_n, \xi_m]$, $\xi_m - \xi_n = (m - n)h$, some modifications are necessary at the endpoints since $2\lfloor n/2 \rfloor$ of the interpolation points at the centers of the relevant B-splines lie outside, to the left and right of D_ξ^n. If the degree is odd (even), these points can be placed at the first and last $\lfloor n/2 \rfloor$ midpoints of knot intervals (knots) in D_ξ^n. The Schoenberg-Whitney conditions hold for both cases which are illustrated in the figure.

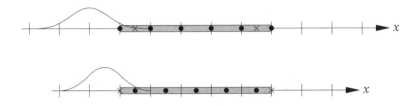

The modifications change some of the first and last rows of the interpolation matrix A. For example, for $n = 3$,

$$A = \frac{1}{48} \begin{pmatrix} 8 & 32 & 8 & & & \\ 1 & 23 & 23 & 1 & & \\ & 8 & 32 & 8 & & \\ & & 8 & 32 & 8 & \\ & & & \ddots & \ddots & \ddots \end{pmatrix}.$$

We notice that the entries do not depend on the grid-width h. This is true in general, and, therefore, tabulated values can be used to generate the interpolation matrices. □

We expect that interpolants provide accurate approximations. This is indeed the case if the norm of the inverse of the interpolation matrix is not too large.

Error of Spline Interpolation

The error of a spline interpolant $p = \sum_{k=0}^{m-1} c_k b_k \in S_\xi^n$ with $\xi : \xi_0, \ldots, \xi_{n+m}$ to a smooth function f can be estimated by

$$|f(x) - p(x)| \leq c\left(n, \|A^{-1}\|_\infty\right) \|f^{(n+1)}\|_{\infty, R} \, h^{n+1}, \quad x \in D_\xi^n = R,$$

where the constant depends on the degree and the maximum norm of the inverse of the interpolation matrix

$$A: a_{i,k} = b_k(x_i),$$

h is the maximal length of the knot intervals, and it is assumed that the parameter interval $D_\xi^n = [\xi_n, \xi_m]$ contains all interpolation points as well as for each B-spline a largest interval in its support.

The coefficients c_k of the spline interpolant p to data $(x_i, g(x_i))$ are computed by solving the linear system

$$\sum_{k=0}^{m-1} a_{i,k} c_k = g(x_i), \quad i = 0, \ldots, m-1.$$

Hence, the interpolation operator can be written in the form

$$g \mapsto Pg = \sum_k \left[\sum_i (A^{-1})_{k,i}\, g(x_i) \right] b_k.$$

Obviously, P is a projector, and the maximum norm of Pg can be estimated by

$$\|Pg\|_{\infty,R} \le \|A^{-1}\|_\infty \|g\|_{\infty,R}$$

since the B-splines sum to one.

To complete the proof, we use the error estimate of a standard projector Q:

$$|f(x) - (Qf)(x)| \le c(n) \|f^{(n+1)}\|_{\infty,R}\, h^{n+1}.$$

Writing

$$f - p = (f - Qf) - P(f - Qf) = g - Pg$$

with $p = Pf$ and $g = f - Qf$, it follows that

$$|f(x) - p(x)| \le \|f - p\|_{\infty,R} \le (1 + \|A^{-1}\|_\infty) \|g\|_{\infty,R},$$

which yields the estimate for the interpolation error.

□ **Example:**

We interpolate with quadratic splines at the knot averages

$$x_i = \xi_i^2 = (\xi_{i+1} + \xi_{i+2})/2,$$

which obviously satisfy the Schoenberg–Whitney conditions. The entries

$$a_{i,k} = b_k(x_i)$$

are nonzero only for $|i - k| \le 1$, and can be computed with the aid of the B-spline recursion. For $k = i + 1$,

$$b_{i+1,\xi}^2(x_i) = \frac{x_i - \xi_{i+1}}{\xi_{i+3} - \xi_{i+1}}\, b_{i+1,\xi}^1(x_i) = \frac{(\xi_{i+1} + \xi_{i+2})/2 - \xi_{i+1}}{(\xi_{i+3} - \xi_{i+2}) + (\xi_{i+2} - \xi_{i+1})}\, \frac{1}{2} = \frac{1}{4}\frac{1}{r_{i+2} + 1}$$

with $r_i = (\xi_{i+1} - \xi_i)/(\xi_i - \xi_{i-1})$. By symmetry,

$$b_{i-1,\xi}^2(x_i) = \frac{1}{4}\frac{1}{1/r_{i+1} + 1},$$

and, since the B-splines sum to one, the ith row of the interpolation matrix equals

$$\ldots, 0, a_{i,i-1} = \frac{1}{4/r_{i+1} + 4}, a_{i,i} = 1 - a_{i,i-1} - a_{i,i+1}, a_{i,i+1}\frac{1}{4r_{i+2} + 4}, 0, \ldots.$$

This formula is valid also if some of the ratios are zero or infinite, e.g., $1/(4/0 + 4) = 0$ and $1/(4/\infty + 4) = 1/4$.

5.6. Smoothing

We bound the norm of A^{-1} for the spline space S_ξ^2 with double knots at the endpoints of the parameter interval and simple interior knots. For this choice of knots, illustrated in the figure, $x_0 = \xi_1 = \xi_2$ and $x_{m-1} = \xi_m = \xi_{m+1}$, so that all interpolation points lie in the parameter interval $D_\xi^2 = [\xi_2, \xi_m]$. Moreover,

$$r = \min_{0 \leq i < m} \max(1/r_{i+1}, r_{i+2}) > 0,$$

since ξ does not contain two consecutive double knots and, hence, in the worst case, one of the ratios $1/r_{i+1}$ or r_{i+2} can be zero. The bound on the local mesh ratios implies an estimate for the off-diagonal entries of the interpolation matrix:

$$a_{i,i-1} + a_{i,i+1} \leq \frac{1}{4} + \frac{1}{4}\frac{1}{1+r} = \frac{1-\varepsilon}{2}.$$

The diagonal entries $a_{i,i}$ are $\geq 1/2$ regardless of the value of r. In view of these inequalities, A is diagonally dominant, and an estimate based on the Neumann series (cf. the appendix) yields the bound

$$\|A^{-1}\|_\infty \leq \frac{2}{\varepsilon} = c(r)$$

for its inverse.

Adapting the general error estimate to the special case under consideration, the quadratic spline interpolant p approximates a smooth function with order 3:

$$|f(x) - p(x)| \leq c(r)\|f^{(3)}\|_{\infty, R} h^3, \quad x \in R,$$

with h the maximal length of the knot intervals in $R = D_\xi^2$. $\square$

5.6 ▪ Smoothing

Cubic spline interpolants at knots have a well-known extremal property discovered by Holladay. In a certain sense, they minimize the size of oscillations among other smooth interpolating functions.

Natural Spline Interpolant

The natural spline interpolant of the data

$$(x_i, f_i), \quad x_0 < x_1 < \cdots < x_M,$$

is a cubic spline p with simple knots at x_ℓ, which satisfies the boundary conditions

$$p''(x_0) = p''(x_M) = 0.$$

Among all twice continuously differentiable interpolants, p minimizes the integral

$$\int_{x_0}^{x_M} |p''(x)|^2 \, dx,$$

which serves as a measure for the oscillations of p.

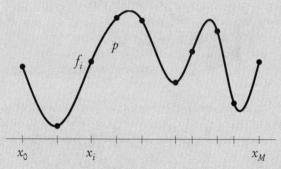

Alternatively, the boundary conditions

$$p'(x_0) = d_0, \quad p'(x_M) = d_M$$

are possible. The resulting clamped natural spline possesses the analogous extremal property.

We first show the existence of an interpolating cubic spline p with the natural boundary conditions. In view of the linearity of the interpolation conditions and since the dimension of the spline space matches the number of data (values and boundary conditions), it is sufficient to show that

$$p(x_0) = \cdots = p(x_M) = 0, \quad p''(x_0) = p''(x_M) = 0$$

implies that p vanishes identically. To this end, we apply Rolle's theorem twice and conclude that there exist $M-1$ interior zeros $x'_1, \ldots, x'_{M-1}$ of p'' with

$$x_{k-1} < x'_k < x_{k+1}.$$

In addition, p'' vanishes at $x'_0 = x_0$ and $x'_M = x_M$. Since

$$p'' = \sum_{k=0}^{M} c_k \, b^1_{k,\xi}$$

with

$$\xi : \xi_0 < \xi_1 = x_0 < \cdots < x_M = \xi_{M+1} < \xi_{M+2}$$

and the linear B-splines $b^1_{k,\xi}$ and the points x'_k satisfy the Schoenberg–Whitney conditions $\xi_k < x'_k < \xi_{k+2}$, it follows that $p'' = 0$. Vanishing at more than one point, p must be zero, too.

5.6. Smoothing

To prove the minimality of p, let f be any twice continuously differentiable function which interpolates the same data:

$$f(x_i) = f_i = p(x_i).$$

Then, we have

$$\int_{x_0}^{x_M} |f''|^2 = \int_{x_0}^{x_M} |(f'' - p'') + p''|^2$$
$$= \int_{x_0}^{x_M} |f'' - p''|^2 + \int_{x_0}^{x_M} |p''|^2 + 2\int_{x_0}^{x_M} (f'' - p'')p''.$$

This identity implies the assertion if we can show that the last integral vanishes. Since $p''(x_0) = 0 = p''(x_M)$, this integral equals

$$-\int_{x_0}^{x_M} (f' - p')p''' = -\sum_{i=1}^{M} c_i \int_{x_{i-1}}^{x_i} f' - p',$$

where c_i denotes the value of the piecewise constant third derivative of the cubic spline p on (x_{i-1}, x_i). Finally, it follows that

$$\int_{x_{i-1}}^{x_i} f' - p' = [f - p]_{x_{i-1}}^{x_i} = 0$$

because f and p interpolate the same data.

The proof of the minimality of the clamped natural spline is completely analogous.

☐ **Example:**

As is illustrated in the figure, we represent the natural cubic spline interpolant in standard form:

$$p = \sum_{k=0}^{m-1} c_k b_k \in S_\xi^3.$$

This means that $\xi_{i+3} = x_i$, $m = M + 3$, $D_\xi^3 = [x_0, x_M]$, and the auxiliary exterior knots are chosen as

$$\xi_{3-k} = x_0 - k(x_1 - x_0), \quad \xi_{m+k} = x_M + k(x_M - x_{M-1})$$

for $k = 1, 2, 3$.

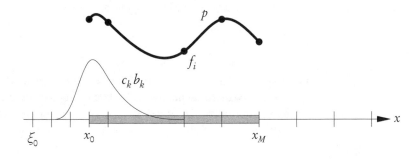

The interpolation conditions yield a linear system

$$Ac = (0, f_0, \ldots, f_M, 0)^t$$

for the B-spline coefficients $(c_0, \ldots, c_{m-1})^t$. The first and last equations correspond to the boundary conditions and the remaining equations to the interpolation conditions $p(x_i) = f_i$.

We determine A explicitly for equally spaced knots

$$\xi_k = \xi_0 + kh.$$

In this case, the values of the B-splines at the interpolation points $x_i, i = 0, \ldots, M$, are

$$b_i(x_i) = \frac{1}{6}, \quad b_{i+1}(x_i) = \frac{2}{3}, \quad b_{i+2}(x_i) = \frac{1}{6},$$

determining the nonzero entries of the rows $2, \ldots, M+2$ of A. It remains to discuss the natural boundary conditions. To this end, we apply the differentiation formula for splines and obtain

$$p'' = h^{-2} \sum_{k=2}^{m-1} c_k'' b_{k,\xi}^1, \quad c_k'' = c_k - 2c_{k-1} + c_{k-2}.$$

Since the values of linear splines at the knots coincide with the coefficients of the corresponding B-spline,

$$p''(x_0) = h^{-2} c_2'', \quad p''(x_M) = h^{-2} c_{m-1}''.$$

Hence, the first and last rows of A equal

$$(1, -2, 1, 0, \ldots), \quad (\ldots, 0, 1, -2, 1),$$

respectively. Since the boundary conditions are homogeneous, the factor h^{-2} is irrelevant.

Summarizing, the linear system has the form

$$\begin{pmatrix} 1 & -2 & 1 & & & 0 \\ 1 & 4 & 1 & & & \\ & \ddots & \ddots & \ddots & & \\ & & 1 & 4 & 1 \\ 0 & & & 1 & -2 & 1 \end{pmatrix} \begin{pmatrix} c_0 \\ c_1 \\ \vdots \\ c_{m-2} \\ c_{m-1} \end{pmatrix} = 6 \begin{pmatrix} 0 \\ f_0 \\ \vdots \\ f_M \\ 0 \end{pmatrix}.$$

For the computation of the clamped natural spline interpolant, we just have to exchange the first and last rows of the linear system. A similar analysis leads to

$$c_2 - c_0 = 2h \, d_0, \quad c_{m-1} - c_{m-3} = 2h \, d_M,$$

where $d_k = f'(x_k)$. $\qquad\square$

Often, function values are perturbed by measurement errors. Matching such noisy data exactly results in unwanted oscillations. A standard remedy, as proposed by Schoenberg and Reinsch, is to minimize a weighted sum of the squares of the interpolation errors in combination with the square integral of the second derivative to control the amount of smoothing.

5.6. Smoothing

Smoothing Spline

The smoothing spline p_σ for the data

$$(x_i, f_i), \quad x_0 < \cdots < x_M,$$

and the weights $w_i > 0$ is the unique cubic spline with simple knots at x_i which minimizes

$$E(p, \sigma) = (1 - \sigma) \sum_{i=0}^{M} w_i |f_i - p(x_i)|^2 + \sigma \int_{x_0}^{x_M} |p''|^2$$

among all twice continuously differentiable functions p.

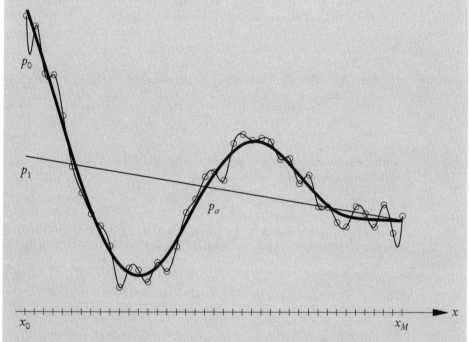

The parameter $\sigma \in (0, 1)$ controls the significance of the data and of the smoothing. For $\sigma \to 0$, p_σ approaches the natural cubic spline interpolant, while, for $\sigma \to 1$, p_σ converges to the least squares line.

To show that the minimizer of E is a cubic spline, we consider a minimizing sequence of twice continuously differentiable functions g_ℓ:

$$E(g_\ell, \sigma) \to \inf_g E(g, \sigma) \geq 0.$$

For any ℓ, the value $E(g_\ell, \sigma)$ can be decreased by replacing g_ℓ by the natural cubic spline interpolant q_ℓ of the data $(x_i, g_\ell(x_i))$. This replacement leaves the sum in the definition of E unchanged, while the integral becomes smaller by the definition of the natural spline as the minimizer of that integral. Hence, $q_0, q_1, \ldots$ is also a minimizing sequence. Since $|q_\ell(x_i)|$ and $\int_D |q_\ell''|^2$ must remain bounded, by compactness and continuity, a subsequence converges to a minimizing natural spline.

The minimizer is unique because of the strict convexity of E. The inequality

$$(\alpha/2 + \beta/2)^2 < (\alpha^2 + \beta^2)/2, \quad \alpha \neq \beta,$$

implies that $E((\varphi + \psi)/2, \sigma) < (E(\varphi, \sigma) + E(\psi, \sigma))/2$ if $\varphi'' \neq \psi''$ or if at least one of the data $\varphi(x_i)$ and $\psi(x_i)$ are different. Hence, if $E(\varphi, \sigma) = E(\psi, \sigma)$ for some $\varphi \neq \psi$, the value of E can be further decreased.

To analyze the behavior of the smoothing spline p_σ as σ tends to zero, we observe that

$$\|q\| = \left(\sum_{i=0}^{M} w_i \, |q(x_i)|^2 \right)^{1/2}$$

is a norm on the space of natural splines, which, by definition, satisfy the boundary conditions $q''(x_0) = q''(x_M) = 0$ and, hence, are uniquely determined by their values at the points x_i. Let us assume that p_σ does not converge to the natural spline q which interpolates the data f_i. This means that there exists $\varepsilon > 0$ such that

$$\|p_\sigma - q\| \geq \varepsilon$$

for a sequence of arbitrarily small σ. By definition of E and since $f_i = q(x_i)$, it follows that

$$E(p_\sigma, \sigma) = (1 - \sigma)\|q - p_\sigma\|^2 + \sigma \int_{x_0}^{x_M} |p_\sigma''|^2 \geq (1 - \sigma)\varepsilon^2 > \sigma \int_{x_0}^{x_M} |q''|^2,$$

with the last inequality valid for small enough σ. Noting that the right side equals $E(q, \sigma)$, this contradicts the minimality of p_σ.

The behavior for $\sigma \to 1$ is discussed in a similar fashion using the seminorm $|q| = \left(\int |q''|^2 \right)^{1/2}$.

We conclude this section by discussing the construction of the smoothing spline. To this end, we represent p_σ in standard form:

$$p_\sigma = \sum_{k=0}^{m-1} c_k \, b_k \in S_\xi^3$$

with $x_i = \xi_{i+3}$, $m = M + 3$, and appropriately chosen exterior knots ξ_0, ξ_1, ξ_2 and ξ_{m+1}, ξ_{m+2}, ξ_{m+3}. To determine p_σ, we express E as a quadratic form in terms of the B-spline coefficients:

$$E(p_\sigma, \sigma) = c^t Q_\sigma c - 2q_\sigma^t c + q_\sigma^0.$$

By solving the linear system

$$Q_\sigma c = q_\sigma,$$

we then obtain the unique minimizing solution.

For assembling the matrix Q_σ and the vector q_σ (the constant q_σ^0 is irrelevant) we consider each part of the quadratic form E in turn. With the $(M + 1) \times m$ interpolation matrix

$$A : a_{i,k} = b_k(x_i)$$

and the diagonal matrix W of weights w_i,

$$\sum_{i=0}^{M} w_i |f_i - p(x_i)|^2 = (f - Ac)^t \, W (f - Ac).$$

The description of the integral term is slightly more complicated. By the differentiation formula,

$$p'' = \sum_{j=0}^{M} d_j\, b^1_{j+2,\xi}, \quad d_j = \alpha^2_{j+2,\xi}\left(\alpha^3_{j+2,\xi}(c_{j+2}-c_{j+1})-\alpha^3_{j+1,\xi}(c_{j+1}-c_j)\right),$$

where $\alpha^n_{k,\xi} = n/(\xi_{k+n}-\xi_k)$. This yields $d = Dc$ with an $(M+1)\times m$ matrix D which contains the appropriate combinations of the factors $\alpha^n_{k,\xi}$. Denoting by

$$G:\; g_{j,j'} = \int_{x_0}^{x_M} b^1_{j+2,\xi}\, b^1_{j'+2,\xi}, \quad j,j' = 0,\ldots,M,$$

the Gramian matrix of the linear B-splines,

$$\int_{x_0}^{x_M} |p''|^2 = \sum_j \sum_{j'} d_j \left(\int b^1_{j+2,\xi}\, b^1_{j'+2,\xi}\right) d_{j'} = c^t D^t G D c.$$

Combining the above formulas and taking the factors $(1-\sigma)$ and σ into account yields

$$Q_\sigma = (1-\sigma)A^t W A + \sigma D^t G D, \quad q_\sigma = (1-\sigma)A^t W f, \quad q^0_\sigma = (1-\sigma)f^t W f,$$

i.e., the desired representation for E.

We could reduce the size of the linear system slightly by using that the smoothing spline p satisfies the natural boundary conditions $p''(x_0) = p''(x_M) = 0$. By the differentiation formula, this implies that

$$\alpha^3_{2,\xi}(c_2 - c_1) - \alpha^3_{1,\xi}(c_1 - c_0) = 0, \quad \alpha^3_{m-1,\xi}(c_{m-1}-c_{m-2}) - \alpha^3_{m-2,\xi}(c_{m-2}-c_{m-3}) = 0.$$

Hence, we can eliminate the first and last unknown coefficients c_0 and c_{m-1}. However, this does not lead to a substantial simplification. Instead, we can view the identities for the coefficients as a test for the correctness of the computed solution.

Chapter 6
Spline Curves

B-spline techniques have become the method of choice for modeling curves. Generalizing the Bézier form, spline representations possess analogous, very intuitive geometric properties. This was realized by Riesenfeld and Gordon, who saw the potential of B-splines for computer aided design. In contrast to polynomials, B-splines allow local shape control. Moreover, they provide approximations of arbitrary accuracy and smoothness.

In Section 6.1, we define polynomial and rational spline curves and, in Section 6.2, discuss basic properties. Then, we describe the fundamental knot insertion and refinement algorithms in Section 6.3. They play a key role for many applications such as conversion to Bézier form and numerical processing of spline curves. Section 6.4 is devoted to evaluation and differentiation. Finally, in Section 6.5, we explain how interpolation schemes for spline functions can be applied to curves.

6.1 ▪ Control Polygon

Spline curves are parametrized by linear combinations of B-splines. The coefficients form a control polygon with geometric properties familiar from the Bézier representation.

> **Spline Curve**
> A spline curve of degree $\leq n$ in $\mathbb{R}^d$ has a parametrization
> $$t \mapsto (p_1(t), \ldots, p_d(t)) = \sum_{k=0}^{m-1} c_k b_k(t), \quad t \in D_\tau^n,$$
> with components p_ν in a spline space S_τ^n with finite knot sequence $\tau: \tau_0, \ldots, \tau_{m+n}$ and parameter interval $D_\tau^n = [\tau_n, \tau_m]$.
>
>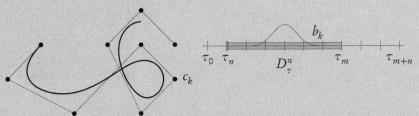
>
> The coefficients $c_k = (c_{k,1}, \ldots, c_{k,d})$ can be combined into an $m \times d$ array C. They are called control points and form the control polygon c for p.

The shape of a spline curve is not only determined by the control points but also influenced by the placement of the knots. This is illustrated in the following example.

☐ **Example:**
The figure shows two cubic spline curves together with their control polygons and knot sequences.

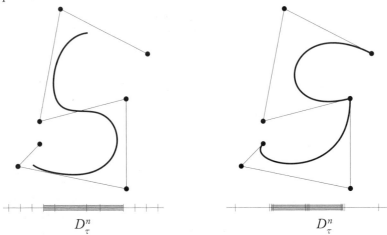

The left curve has simple knots; hence, the parametrization is twice continuously differentiable. The example on the right shows how nonsmooth curves can be modeled with the aid of multiple knots. At a knot with multiplicity n, the derivative of the parametrization need not be continuous. Hence, the tangent direction can change abruptly. Moreover, due to the n-fold knots at the endpoints of the parameter interval, the curve interpolates the first and the last control points.   ☐

If the spline parametrization is periodic, the corresponding curve is closed. Accordingly, the control polygon is also closed, reflecting the periodicity of the B-spline coefficients. A precise definition, also introducing the proper notation, is given below.

Closed Spline Curve
A closed spline curve of degree $\leq n$ in $\mathbb{R}^d$ has a parametrization
$$t \mapsto (p_1(t),\ldots,p_d(t)) = \sum_{k \in \mathbb{Z}} c_k\, b_k(t), \quad t \in \mathbb{R},$$
with components which are continuous T-periodic splines. This means that $p_\nu \in S^n_{\tau,T}$, $\tau = (\tau_0,\ldots,\tau_{M-1})$, and the B-splines b_k correspond to the periodically extended knot sequence $(\ldots,\tau-T,\tau,\tau+T,\ldots)$ and to M-periodic control points.

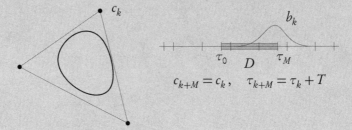

6.1. Control Polygon

> According to the periodicity conditions, p is determined by M consecutive control points
> $$C = \begin{pmatrix} c_0 \\ \vdots \\ c_{M-1} \end{pmatrix},$$
> which form a closed control polygon for p.

The following example illustrates the representation of closed spline curves.

☐ **Example:**
The figure shows two closed cubic spline curves together with their knots $(\tau_0, \ldots, \tau_{M-1})$ in a periodicity interval $D = [\tau_0, \tau_0 + T)$.

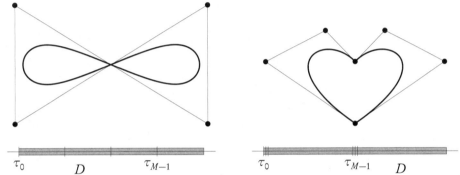

We note that the number of control points exactly corresponds to the number of knots in the half-open periodicity interval. For the example on the right, each knot has multiplicity 3. In particular, $\tau_0 = \tau_1 = \tau_2$ and $\tau_M = \tau_{M+1} = \tau_{M+2}$ are the two endpoints of D. ☐

As is illustrated in the figure below, we can describe a closed spline curve $p(t)$, $t \in D = [\tau_0, \tau_0 + T]$, with a nonperiodic parametrization. We adjoin to the knots $\tau_0, \ldots, \tau_{M-1}$, $\tau_M = \tau_0 + T$, successively n knots on each side of D according to the periodicity condition:
$$\tau \to \tilde{\tau} : (\tau_{M-n} - T, \ldots, \tau_{M-1} - T, \tau_0, \ldots, \tau_{M-1}, \tau_0 + T, \ldots, \tau_n + T).$$

The corresponding control points are
$$c_{M-n}, \ldots, c_{M-1}, c_0, \ldots, c_{M-1},$$
yielding a parametrization with components in the spline space $S_{\tilde{\tau}}^n$. We note that for $n > M$ the auxiliary control points c_k have to be defined successively, for $k = -1, -2, \ldots$.

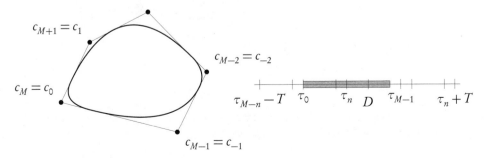

The alternative representation of a closed spline curve is convenient for numerical methods since algorithms are usually formulated for the nonperiodic case. With the above identification, closed curves can be handled with minor modifications only.

As for Bézier curves, we can use rational parametrizations. While the gain in accuracy and design flexibility is marginal, a number of standard shapes can be represented exactly. Therefore, this most general curve format has become a standard in many modeling systems.

Rational Parametrizations

A nonuniform rational B-spline parametrization (NURBS) $r = p/q$ is the quotient of a spline parametrization $t \mapsto p(t)$ with weighted control points

$$(c_k w_k) \in \mathbb{R}^d, \quad w_k > 0,$$

and a spline function $t \mapsto q(t)$ with coefficients w_k.

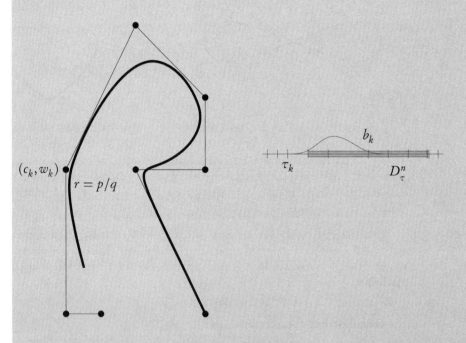

The weights w_k provide additional design flexibility. Increasing a weight pulls the curve towards the corresponding control point.

We can identify r with a spline curve in homogeneous coordinates parametrized by

$$t \mapsto (p(t) | q(t)) = \sum_k (c_k w_k | w_k) b_k(t) \in \mathbb{R}^{d+1}.$$

This interpretation is convenient for implementing algorithms such as knot insertion, evaluation, and differentiation.

As an example, we discuss the relationship to the rational Bézier form which is a special case of the NURBS representation.

6.1. Control Polygon

□ **Example:**

Conic sections can be represented by rational quadratic Bézier curves. For example, the quarter circle with the standard trigonometric parametrization

$$(\cos\varphi, \sin\varphi), \quad \varphi \in [0, \pi/2],$$

has the Bézier control points and weights

$$(c_k | w_k): \quad (1,0|1), (1,1|1/\sqrt{2}), (0,1|1).$$

Accordingly, a full circle can be represented by a closed rational quadratic spline curve r in Bézier form with double knots

$$\tau_0 = 0, 0, 1, 1, 2, 2, 3, 3 = \tau_7, \quad \tau_{k+8} = \tau_k + 4$$

and control points and weights shown in the figure.

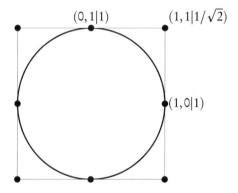

It is perhaps a bit surprising that the Bézier form does not correspond to a closed rational quadratic spline curve with simple uniform knots parametrized by

$$t \mapsto r(t) = \frac{\sum_k (c_k w_k) b^2(t-k)}{\sum_k w_k b^2(t-k)},$$

where

$$c_0 = (1,1), \, c_1 = (-1,1), \, c_2 = (-1,-1), \, c_3 = (1,-1), \quad c_{k+4} = c_k.$$

If this were the case, $b^2(1) = b^2(2) = 1/2$ implies

$$1 = |r(k)|^2 = \left| \frac{c_{k-2} w_{k-2}/2 + c_{k-1} w_{k-1}/2}{w_{k-2}/2 + w_{k-1}/2} \right|^2.$$

In view of the specific form of the control points this yields

$$(w_{k-2} + w_{k-1})^2 + (w_{k-2} - w_{k-1})^2 = (w_{k-2} + w_{k-1})^2.$$

We conclude that all weights must be equal. But then the parametrization is polynomial which is not possible. □

6.2 ▪ Basic Properties

The geometric properties of the Bézier control polygon persist for spline curves. In particular, with appropriate choice of the knots, the endpoint interpolation property can be maintained.

Properties of the Control Polygon
The control polygon c of a spline curve parametrized by

$$p = \sum_{k=0}^{m-1} c_k b_k, \quad p_\nu \in S_\tau^n, \quad \tau: \tau_0, \ldots, \tau_{m+n},$$

qualitatively models the shape of p.

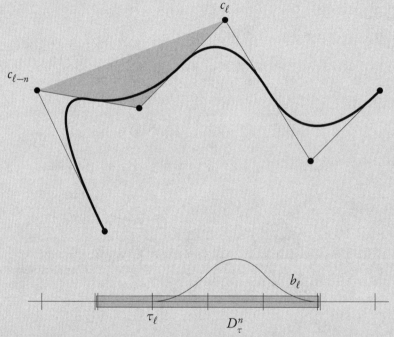

As is illustrated by the figure,

- for $\tau_\ell \leq t \leq \tau_{\ell+1}$ the point $p(t)$ lies in the convex hull of $c_{\ell-n}, \ldots, c_\ell$.

Moreover, if both endpoints of the parameter interval $D_\tau^n = [\tau_n, \tau_m]$ are knots with multiplicity n, then

- $p(\tau_n) = c_0, \ p(\tau_m) = c_{m-1}$,
- $p'(\tau_n^+) = \alpha_{1,\tau}^n (c_1 - c_0), \ p'(\tau_m^-) = \alpha_{m-1,\tau}^n (c_{m-1} - c_{m-2})$

with $\alpha_{k,\tau}^n = n/(\tau_{k+n} - \tau_k)$. The last two properties referred to as endpoint interpolation imply that the control polygon is tangent to the spline curve, which is very useful for design purposes.

While the parametrization of a spline curve is continuous, the derivative can have jumps. This is taken into account in the formula for p', where the superscripts $+/-$ denote limits from the right/left.

6.2. Basic Properties

The properties are easily verified.

Since only the B-splines b_k, $k = \ell - n, \ldots, \ell$, are relevant for $t \in [\tau_\ell, \tau_{\ell+1}]$, and they are positive and sum to one, the convex hull property is obvious.

To check the endpoint interpolation, we consider, e.g., the left endpoint. Because of the knot multiplicity n, all B-splines except b_0 vanish at τ_n. Moreover, $b_0(\tau_n) = 1$, so that $p(\tau_n)$ coincides with the control point c_0. By the differentiation formula,

$$p' = \sum_{k=1}^{n} \alpha_{k,\tau}^n (c_k - c_{k-1}) b_{k,\tau}^{n-1}, \quad \alpha_{k,\tau}^n = n/(\tau_{k+n} - \tau_k),$$

on the interval (τ_n, τ_{n+1}). Since

$$b_{k,\tau}^{n-1}(\tau_n^+) = \delta_{1,k},$$

the formula for $p'(\tau_n^+)$ follows.

□ **Example:**

The figure illustrates the convex hull property and endpoint interpolation for spline curves of degree 2 and 3.

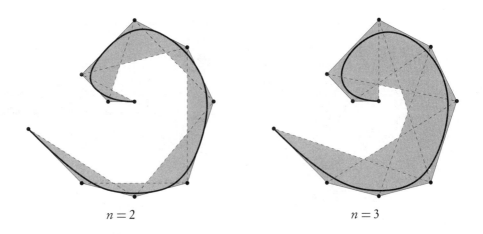

$n = 2$ $\qquad\qquad\qquad\qquad$ $n = 3$

Obviously, for small degree, the convex hull provides a more accurate approximation. Moreover, if the control polygon contains nearly flat parts, the bounds are very tight. □

The convex hull property implies that the control polygon c is close to the corresponding curve p. In particular, if c is linear, so is p. Therefore, one expects that the distance to the spline curve is related to the second differences of the control points as a measure for the deviation from linearity. This is indeed the case as implied by the following precise bound due to Reif.

Distance to the Control Polygon
For a spline curve parametrized by

$$p = \sum_{k=0}^{m-1} c_k b_k, \quad p_\nu \in S_\tau^n,$$

with $n > 1$, let c be a piecewise linear parametrization of the control polygon, which interpolates c_k at the knot averages $\tau_k^n = (\tau_{k+1} + \cdots + \tau_{k+n})/n$.

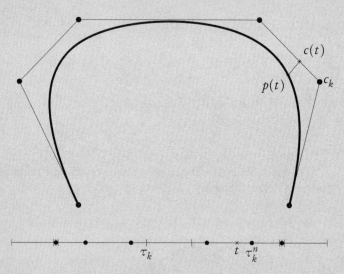

The distance of p to the control polygon can be bounded in terms of weighted second differences of the control points. For t in a nondegenerate knot interval $[\tau_\ell, \tau_{\ell+1}] \subseteq D_\tau^n$,

$$\|p(t) - c(t)\|_\infty \le \frac{1}{2n} \max_{\ell-n \le k \le \ell} \sigma_k^2 \max_{\ell-n+2 \le k \le \ell} \|\nabla_\tau^2 c_k\|_\infty,$$

where

$$\sigma_k^2 = \frac{1}{n-1} \sum_{i=1}^{n} (\tau_{k+i} - \tau_k^n)^2$$

and $\nabla_\tau^2 c_k$ are the control points of the second derivative p''. Their explicit form is

$$\nabla_\tau^2 c_k = \frac{n-1}{\tau_{k+n-1} - \tau_k} \left(\frac{c_k - c_{k-1}}{\tau_k^n - \tau_{k-1}^n} - \frac{c_{k-1} - c_{k-2}}{\tau_{k-1}^n - \tau_{k-2}^n} \right),$$

where none of the denominators vanishes since the differences are at least as large as $\tau_{\ell+1} - \tau_\ell$.

The local estimate implies a global bound simply by taking the maximum of the right side over all k relevant for the entire parameter interval $D_\tau^n = [\tau_n, \tau_m]$ of the spline curve. In this case, $\nabla_\tau^2 c_k$ is set to 0 if $\tau_k = \tau_{k+n-1}$.

6.2. Basic Properties

Clearly, it suffices to prove the estimate for each component separately. Therefore, we may consider the univariate case. By scaling the coefficients c_k, which appear linearly on both sides of the estimate, we can further assume that $\max_k |\nabla_\tau^2 c_k| = 1$.

Turning to the main arguments of the proof, we first show that the worst case corresponds to a parabola p^*,

$$p^*(t) = t^2/2 = \sum_k c_k^* b_k, \quad \nabla_\tau^2 c_k^* = 1.$$

In terms of the parametrizations $c(t)$ and $c^*(t)$ of the control polygons, this means that

$$-(p^* - c^*) \le p - c \le p^* - c^*$$

for any other spline $p = \sum_k c_k b_k$ with $\max_k |\nabla_\tau^2 c_k| = 1$.

Both inequalities follow from a simple geometric observation. Considering, e.g., an equivalent form of the right estimate

$$p - p^* = \sum_k (c_k - c_k^*) b_k \le c - c^*,$$

we observe that $\nabla_\tau^2 (c_k - c_k^*) \le 0$ in view of our assumption about the size of the weighted second differences. By definition, ∇_τ^2 forms differences of consecutive slopes of the polygon, which connects the points $(\tau_k^n, c_k - c_k^*)$ weighted by the positive factors $(n-1)/(\tau_{k+n-1} - \tau_k)$. Hence, the polygon is concave. Since, by Marsden's identity,

$$(t, p(t) - p^*(t)) = \sum_k (\tau_k^n, c_k - c_k^*) b_k(t),$$

the convex hull property implies that the graph of the spline $p - p^*$ lies below the polygon $c - c^*$, as claimed.

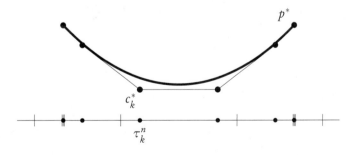

It remains to bound $p^* - c^*$, which then yields the estimate for $|p - c|$. Since the second derivative of this function, which is the difference between the parabola and the broken line shown in the figure, equals one on each interval (τ_k^n, τ_{k+1}^n), it follows that its maximum must occur at one of the knot averages. With the aid of Marsden's identity we can compute the explicit form of the control points for $p^*(t) = t^2/2$:

$$c_k^* = (\tau_k^n)^2/2 - \sigma_k^2/(2n)$$

(cf. the example in Section 4.3). Substituting this expression into
$$p^*(\tau_k^n) - c^*(\tau_k^n) = (\tau_k^n)^2/2 - c_k^*,$$
the desired estimate follows; i.e., we can bound $|p - c|$ by the maximum of $\sigma_k^2/(2n)$ for the relevant indices k.

☐ **Example:**
As an example, we consider the quadratic spline curve $p = \sum_{k=0}^{4} c_k b_k$ shown in the left figure with
$$C = \begin{pmatrix} 1 & 0 & 0 & 1 & 2 \\ 0 & 0 & 1 & 2 & 2 \end{pmatrix}^t, \quad \tau = (-1, 0, 0, 2, 6, 12, 14, 15).$$
The right diagram shows the error $e(t) = \|p(t) - c(t)\|_\infty$ for t in the parameter interval $D_\tau^2 = [0, 12]$ as well as the piecewise constant upper bound provided by the estimate.

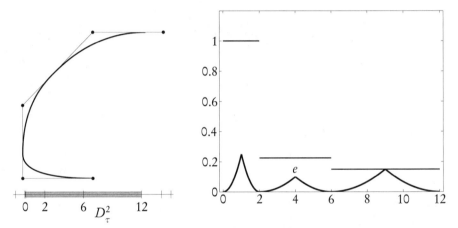

Specializing the general formulas for degree $n = 2$, we obtain
$$\tau_k^2 = (\tau_{k+1} + \tau_{k+2})/2, \quad \sigma_k^2 = (\tau_{k+2} - \tau_{k+1})^2/2.$$
The concrete values are
$$(\tau_0^2, \sigma_0^2) = (0,0), (1,2), (4,8), (9,18), (13,2) = (\tau_4^2, \sigma_4^2).$$
Finally, the weighted control point differences equal
$$\nabla_\tau c_k = \frac{c_k - c_{k-1}}{\tau_k^2 - \tau_{k-1}^2} : \quad (-1,0), (0,1/3), (1/5,1/5), (1/4,0),$$
and
$$\nabla_\tau^2 c_k = \frac{\nabla_\tau c_k - \nabla_\tau c_{k-1}}{\tau_{k+1} - \tau_k} : \quad (1/2, 1/6), (1/20, -1/30), (1/120, -1/30)$$
with the index k starting from 1 and 2, respectively. This leads to the following estimates. For $t \in [0, 2]$,
$$\|p(t) - c(t)\|_\infty \leq \frac{1}{4} \cdot \max\{0, 2, 8\} \cdot \frac{1}{2} = 1.$$

Similarly, for the other two knot intervals $[2, 6]$ and $[6, 12]$, the bounds are

$$\frac{1}{4} \cdot 18 \cdot \frac{1}{20} = \frac{9}{40}, \quad \frac{1}{4} \cdot 18 \cdot \frac{1}{30} = \frac{3}{20},$$

respectively.

The right figure shows that the estimate for the last knot interval is sharp. □

As a second example, we consider the case of uniform knots which permits slight simplifications.

□ **Example:**

For uniform knots τ_k with grid width h and odd degree $n = 2m + 1$,

$$\tau_k^n = \tau_k + (m+1)h, \quad \sigma_k^2 = \frac{1}{n-1} \sum_{i=-m}^{m} (ih)^2.$$

Hence, the product of the first two factors in the error estimate for the control polygon equals

$$\frac{1}{2n} \frac{2h^2}{(n-1)} \sum_{i=1}^{m} i^2.$$

Substituting $m^3/3 + m^2/2 + m/6$ for the sum and cancelling common factors, we obtain the simplified expression

$$\frac{n+1}{24} h^2.$$

Therefore, since

$$\nabla_\tau^2 c_k = h^{-2} \Delta^2 c_{k-2}, \quad \Delta^2 c_{k-2} = c_k - 2c_{k-1} + c_{k-2},$$

the error estimate for uniform knots has the form

$$\|p(t) - c(t)\|_\infty \leq \frac{n+1}{24} \max_{\ell-n \leq k \leq \ell-2} \|\Delta^2 c_k\|_\infty$$

for $\tau_\ell \leq t < \tau_{\ell+1}$.

An analogous computation shows that the same formula also holds for even degree $n = 2m$. The fact that in this case the knot averages do not coincide with knots but are midpoints of knot intervals does not make any difference. □

6.3 ▪ Refinement

Refinement of the partition of the parameter interval is the standard procedure to increase the local accuracy or design flexibility of spline approximations. For spline curves, the corresponding algorithm for inserting new knots has a simple geometric interpretation in terms of the control polygon, as observed, independently, by Boehm and Cohen, Lyche, and Riesenfeld. It plays a fundamental role also for other applications.

Knot Insertion

Let $p = \sum_0^{m-1} c_k b_k \in S_\tau^n$ parametrize a spline curve. If we add a new knot s in the parameter interval D_τ^n and $s \in [\tau_\ell, \tau_{\ell+1})$, then the control points $\tilde{c}_k$ of p with respect to the refined knot vector

$$\tilde{\tau}: \quad \ldots, \tilde{\tau}_\ell = \tau_\ell, \tilde{\tau}_{\ell+1} = s, \tilde{\tau}_{\ell+2} = \tau_{\ell+1}, \ldots$$

are computed as follows.

On the segments $[c_{k-1}, c_k]$ with $\tau_k < s < \tau_{k+n}$ new control points are generated:

$$\tilde{c}_k = \gamma_{k,\tau}^n c_k + (1 - \gamma_{k,\tau}^n) c_{k-1}, \quad \gamma_{k,\tau}^n = \frac{s - \tau_k}{\tau_{k+n} - \tau_k}.$$

The other edges of the control polygon remain unchanged:

$$\tilde{c}_k = c_k \text{ for } \tau_{k+n} \leq s, \quad \tilde{c}_k = c_{k-1} \text{ for } s \leq \tau_k.$$

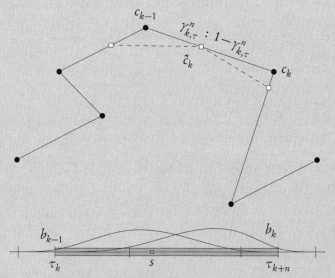

As illustrated in the figure, the new control point $\tilde{c}_k$ divides the segment $[c_{k-1}, c_k]$ in the same ratio as the parameter s divides the interval $[\tau_k, \tau_{k+n}]$, which is the intersection of the supports of the two associated B-splines. We note that, if s coincides with a knot ($s = \tau_\ell$), fewer control points need to be computed. More precisely, if s has multiplicity j in $\tilde{\tau}$, only $n + 1 - j$ convex combinations need to be formed.

Several new knots can be inserted by repeating the procedure. In particular, by raising the multiplicity of a knot to n, we obtain a point on the curve:

$$\tau_{\ell-n} < \tau_{\ell-n+1} = \cdots = \tau_\ell < \tau_{\ell+1} \implies p(\tau_\ell) = c_{\ell-n}.$$

Hence, the evaluation scheme for splines can be viewed as n-fold knot insertion.

To verify the knot insertion formula, we have to show that

$$\sum_k c_k b_k(t) = \sum_k \tilde{c}_k \tilde{b}_k(t), \quad \tilde{c}_k = \tilde{\gamma}_k c_k + (1 - \tilde{\gamma}_k) c_{k-1},$$

where $\tilde{b}_k$ denote the B-splines of the refined knot sequence $\tilde{\tau}$ and

$$\tilde{\gamma}_k = \begin{cases} 1 & \text{if } \tau_{k+n} \leq s, \\ \gamma_k & \text{if } \tau_k < s < \tau_{k+n}, \\ 0 & \text{if } s \leq \tau_k. \end{cases}$$

The modified weights $\tilde{\gamma}_k$ incorporate the cases where the control points do not change.

As usual, for a proof, it is convenient to extend τ and $\tilde{\tau}$ to bi-infinite knot sequences and to restrict t to the parameter interval $D_\tau^n = [\tau_n, \tau_m]$ of the curve. Moreover, since the modified weights depend continuously on the knots, it is sufficient to consider simple knots.

With these notations and assumptions, the identity can be derived from general considerations. Since adding a knot enlarges the class of splines, any B-spline b_k can be written as a linear combination of the B-splines corresponding to $\tilde{\tau}$:

$$b_k = \sum_i d_{k,i} \tilde{b}_i.$$

Moreover, since b_k vanishes outside of the interval $[\tilde{\tau}_k, \tilde{\tau}_{k+n+2})$, it follows from the local linear independence of the B-splines that the coefficients can be nonzero only for $i = k$ and $i = k+1$. These observations imply that

$$\sum_k c_k b_k = \sum_k c_k (p_k \tilde{b}_k + q_k \tilde{b}_{k+1}) = \sum_k (c_k p_k + c_{k-1} q_{k-1}) \tilde{b}_k.$$

Choosing $c_k = 1$, we obtain in particular

$$q_{k-1} = 1 - p_k$$

since the B-splines form a partition of unity.

It remains to determine the explicit form of the weight p_k. This can be done by analyzing the linear combination representing a single B-spline, which takes on the form

$$b_k = p_k \tilde{b}_k + q_k \tilde{b}_{k+1}.$$

We consider several cases.

(i) If $\tau_{k+n} \leq s$, then b_{k-1} and $\tilde{b}_{k-1}$ have the same knots. Consequently, $b_{k-1} = \tilde{b}_{k-1}$, so that

$$1 = p_{k-1}, \quad 0 = q_{k-1} = 1 - p_k.$$

(ii) If $\tau_k < s < \tau_{k+n}$, then b_k has the knots $\tau_k, \dots, \tau_{k+n+1}$, while $\tilde{b}_k$ has the knots $\tau_k, \dots, s, \dots, \tau_{k+n}$. It follows from the recursion for B-splines that

$$b_k(t) = \frac{t - \tau_k}{\tau_{k+1} - \tau_k} \cdots \frac{t - \tau_k}{\tau_{k+n} - \tau_k},$$
$$\tilde{b}_k(t) = \frac{t - \tau_k}{\tau_{k+1} - \tau_k} \cdots \frac{t - \tau_k}{s - \tau_k} \cdots \frac{t - \tau_k}{\tau_{k+n-1} - \tau_k}$$

for $t \in [\tau_k, \min(\tau_{k+1}, s))$. Since the B-spline $\tilde{b}_{k+1}$ vanishes on this interval,

$$p_k = \frac{s - \tau_k}{\tau_{k+n} - \tau_k} = \gamma_{k,\tau}^n.$$

(iii) If $s \leq \tau_k$, then b_k and $\tilde{b}_{k+1}$ have the same knots. Hence, $b_k = \tilde{b}_{k+1}$ and
$$1 = q_k = 1 - p_{k+1}, \quad 0 = p_k.$$

In all cases,
$$p_k = \tilde{\gamma}_k,$$

proving the knot insertion formula.

☐ **Example:**

We illustrate the knot insertion algorithm for the cubic spline curve $\sum_{k=0}^{6} c_k \, b_k$ with control points
$$C = \begin{pmatrix} 0 & 4 & 4 & 1 & 1 & 5 & 4 \\ 2 & 1 & 4 & 4 & 9 & 8 & 6 \end{pmatrix}^t$$

and knot sequence
$$\tau_0 = 1, 2, 2, 2, 3, 5, 5, 8, 9, 10, 11 = \tau_{10},$$

shown in the figure.

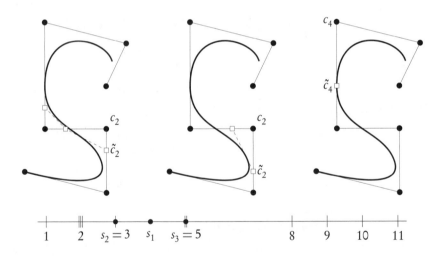

We consider three cases.

(i) $s = 4$ (multiplicity 1): Three segments of the control polygon are subdivided. The ratios
$$2:1, 2:1, 1:4$$

are determined by the relative position of s in the intervals
$$[\tau_k, \tau_{k+3}]: [2,5], [2,5], [3,8],$$

which are the intersections of the supports of the relevant B-splines. The resulting control points, corresponding to the refined knot sequence
$$\tilde{\tau}_0 = 1, 2, 2, 2, 3, 4, 5, 5, 8, 9, 10, 11 = \tilde{\tau}_{11},$$

6.3. Refinement

are
$$\tilde{C} = \begin{pmatrix} 0 & 4 & 4 & 2 & 1 & 1 & 5 & 4 \\ 2 & 1 & 3 & 4 & 5 & 9 & 8 & 6 \end{pmatrix}^t.$$

For example,
$$(4,3) = \tilde{c}_2 = \frac{2}{3} c_2 + \frac{1}{3} c_1$$

(cf. the left figure).

(ii) $s = 3$ (double knot): In this case, only the second and third segments are subdivided according to the ratios
$$1:2, \; 1:2,$$
as is shown in the middle figure.

(iii) $s = 5$ (triple knot): As is shown in the right figure, if the multiplicity after knot insertion is maximal, only one new control point is generated, which lies on the curve. In this example,
$$p(s) = \tilde{c}_4 = \frac{2}{5}(1,9) + \frac{3}{5}(1,4) = (1,6). \quad \square$$

Instead of inserting one knot at a time, several knots can be inserted simultaneously. For example, all knot intervals can be split in half to gain more approximation power on the entire parameter interval. The corresponding algorithm has a particularly elegant form for spline curves with uniform knots as was described by Lane and Riesenfeld.

Uniform Subdivision

Let
$$p = \sum_k c_k b_{k,\tau}^n \in S_\tau^n$$

be a parametrization of a uniform spline curve with grid width h. If new knots are simultaneously inserted at the midpoints of the knot intervals, the control points $\tilde{c}_k$ corresponding to the refined knot sequence $\tilde{\tau}$ with grid width $h/2$ and $\tilde{\tau}_{2k} = \tau_k$ can be computed as follows.

(i) The control points c_k, $k \sim \tau$, are doubled:
$$\tilde{c}_{2k} = \tilde{c}_{2k+1} = c_k.$$

(ii) Simultaneous averages of adjacent control points are formed,
$$\tilde{c}_k \leftarrow (\tilde{c}_k + \tilde{c}_{k-1})/2,$$

and this process is repeated n-times.

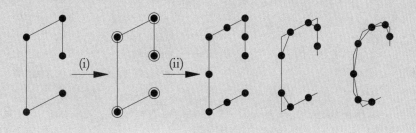

The explicit form of the new control points is

$$\tilde{c}_k = \sum_i s_{k-2i}\, c_i, \quad s_j = 2^{-n}\binom{n+1}{j},$$

where $s_j = 0$ for $j < 0$ or $j > n+1$ in accordance with the convention for binomial coefficients.

We note that for a finite knot sequence $\tau_0,\dots,\tau_{m+n}$ the indices of the control points $\tilde{c}_k$ are not in the standard range $\{0,1,\dots\}$. By definition of $\tilde{\tau}$, the B-splines $b^n_{k,\tilde{\tau}}$, $k = n,\dots,2m-1$, are relevant for the parameter interval D^n_τ. We could have changed the definition of $\tilde{\tau}$. This would, however, result in less elegant subdivision formulas.

Turning to the derivation of the subdivision formula, it is sufficient to consider spline functions since the curve components are processed with identical operations. Moreover, in view of linearity, we need to check the identity only for a single B-spline b_ℓ. In this case, the coefficients are $c_k = \delta_{\ell-k}$ and the formula reduces to

$$\tilde{c}_k = s_{k-2\ell} = 2^{-n}\binom{n+1}{k-2\ell},$$

where $\tilde{c}_k$ are the coefficients for representing b_ℓ as a linear combination of the B-splines $\tilde{b}_k$ of the refined knot sequence $\tilde{\tau}$.

Using induction on the degree n, it is convenient to start with $n = 0$, although this case is excluded for splines. Since the B-splines of degree 0 are the characteristic functions of the half-open knot intervals and $\tilde{\tau}_{2\ell} = \tau_\ell$,

$$b^0_{\ell,\tau} = \tilde{b}^0_{2\ell,\tau} + \tilde{b}^0_{2\ell+1,\tau}.$$

This is in agreement with

$$\tilde{c}_{2\ell} = \tilde{c}_{2\ell+1} = 1, \quad \tilde{c}_k = 0,\, k \notin \{2\ell, 2\ell+1\},$$

as asserted by the subdivision formula.

For the induction step $n-1 \to n$, we have to show that

$$b^n_{\ell,\tau} = 2^{-n}\sum_k \binom{n+1}{k-2\ell} b^n_{k,\tilde{\tau}},$$

recalling that we are considering the special case $c_k = \delta_{\ell-k}$; i.e., the spline function $\sum_k c_k b^n_{k,\tau}$ is the single B-spline $b^n_{\ell,\tau}$. We differentiate the formula for subdividing $b^n_{\ell,\tau}$, using the simple recursion for the derivative of uniform B-splines. Since both sides vanish at $\tau_\ell = \tilde{\tau}_{2\ell}$, this yields the equivalent identity

$$B_\ell - B_{\ell+1} = 2^{1-n}\underbrace{\left[\sum_k \binom{n+1}{k-2\ell}(\tilde{B}_k - \tilde{B}_{k+1})\right]}_{S_1},$$

where we have set $B_k = b^{n-1}_{k,\tau}$ and $\tilde{B}_k = b^{n-1}_{k,\tilde{\tau}}$ to simplify notation. The factor 2^{-n} has changed since the differentiation formula involves the different factors h^{-1} and $(h/2)^{-1}$ for the B-splines on the left and right side, respectively. We now use the induction hypothesis and express the B-splines on the left side in terms of the B-splines $\tilde{B}_k$ corresponding to $\tilde{\tau}$. Noting that

$$B_{\ell+1}(x) = B_\ell(x-h), \quad \tilde{B}_k(x-h) = \tilde{B}_{k+2}(x),$$

this yields

$$2^{-(n-1)} \underbrace{\left[\sum_k \binom{n}{k-2\ell}(\tilde{B}_k - \tilde{B}_{k+2})\right]}_{S_2}.$$

In both expressions $S_\nu = [\dots]$, we sum over $k \in \mathbb{Z}$, using the convention that $\binom{m}{j} = 0$ for $j < 0$ or $j > m$. The equality of these two sums follows by comparing the coefficients of $\tilde{B}_j$, which are

$$\binom{n+1}{j-2\ell} - \binom{n+1}{j-1-2\ell}$$

for S_1 and

$$\binom{n}{j-2\ell} - \binom{n}{j-2-2\ell}$$

for S_2. The expressions agree in view of the recursion for binomial coefficients:

$$\binom{n+1}{j-2\ell} = \binom{n}{j-2\ell} + \binom{n}{j-2\ell-1}, \quad \binom{n+1}{j-1-2\ell} = \binom{n}{j-1-2\ell} + \binom{n}{j-2-2\ell}.$$

It remains to derive the algorithmic definition of the coefficients $\tilde{c}_k$. Again, we can consider spline functions and make use of linearity. Hence, assuming that $c_k = \delta_{k-\ell}$, step (i) of the algorithm yields

$$\dots, \tilde{c}_{2\ell} = 1, 1 = \tilde{c}_{2\ell+1}, \dots,$$

where we list only nonzero coefficients. Applying step (ii) repeatedly, we obtain

$$\tilde{c}_{2\ell} = \begin{matrix} 1/2, & 2/2, & 1/2 \\ 1/4, & 3/4, & 3/4, & 1/4 \\ 1/8, & 4/8, & 6/8, & 4/8, & 1/8 \\ & \dots & \end{matrix}$$

We recognize Pascal's triangle, i.e., the algorithm leads to the coefficients $2^{-n}\binom{n+1}{k-2\ell}$, as asserted by the subdivision formula.

□ **Example:**
We apply the subdivision algorithm to the cubic closed spline curve with control points at the vertices $(\pm 1, \pm 1)$ of a square with side length 2.

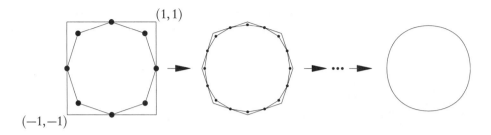

In this case, the subdivision rule is

$$\begin{array}{rcccl}\tilde{c}_{2\ell} &=& s_4 c_{\ell-2} + s_2 c_{\ell-1} + s_0 c_\ell &=& \frac{1}{8} c_{\ell-2} + \frac{3}{4} c_{\ell-1} + \frac{1}{8} c_\ell, \\ \tilde{c}_{2\ell+1} &=& s_3 c_{\ell-1} + s_1 c_\ell &=& \frac{1}{2} c_{\ell-1} + \frac{1}{2} c_\ell.\end{array}$$

Two steps of the corresponding refinements as well as the spline curve obtained in the limit are shown in the figure.

To analyze the iteration a little bit further, we derive an explicit formula for a control point sequence, emerging from one of the corners. To this end, we write the modification of three adjacent control points in matrix form:

$$\begin{pmatrix} \tilde{c}_{2\ell-1} \\ \tilde{c}_{2\ell} \\ \tilde{c}_{2\ell+1} \end{pmatrix} = \underbrace{\begin{pmatrix} 1/2 & 1/2 & 0 \\ 1/8 & 3/4 & 1/8 \\ 0 & 1/2 & 1/2 \end{pmatrix}}_{A} \begin{pmatrix} c_{\ell-2} \\ c_{\ell-1} \\ c_\ell \end{pmatrix}.$$

The subdivision matrix has the eigenvalues 1, $1/2$, and $1/4$, and, by transformation to diagonal form, it can be shown that

$$A_\infty = \lim_{\ell \to \infty} A^\ell = \frac{1}{6} \begin{pmatrix} 1 & 4 & 1 \\ 1 & 4 & 1 \\ 1 & 4 & 1 \end{pmatrix}.$$

This proves, e.g., that the sequence of control points emerging from $(1,1)$ tends to the limit

$$\frac{1}{6} \begin{pmatrix} 1 & 4 & 1 \end{pmatrix} \begin{pmatrix} -1 & 1 \\ 1 & 1 \\ 1 & -1 \end{pmatrix} = \frac{2}{3} \begin{pmatrix} 1 & 1 \end{pmatrix},$$

which is a point on the curve. □

As a more theoretical application of knot insertion, we discuss a measure for the number of oscillations of a spline curve.

> **Variation Diminution**
> The variation of a spline curve parametrized by $p = \sum_{k=0}^{m-1} c_k b_k$ with respect to a hyperplane H is not larger than the variation of its control polygon c:
>
> $$V(p, H) \leq V(c, H),$$
>
> where V denotes the maximal number of pairs of consecutive points on opposite sides of H.

6.3. Refinement

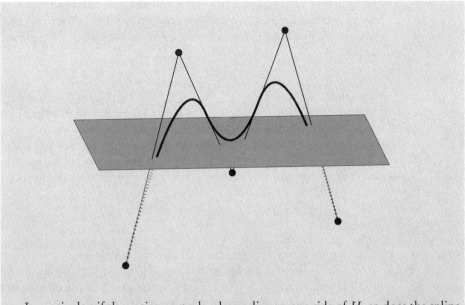

In particular, if the entire control polygon lies on one side of H, so does the spline curve.

This important property of spline curves can be proved by elegant geometric arguments, as was observed by Lane and Riesenfeld.

Since the spline curve consists of finitely many polynomial segments, its variation is finite. Hence, in the nontrivial case, when $V(p,H) > 0$, there exist pairs of points

$$p(t_{k-1}),\ p(t_k),\quad k = 1,\ldots,V(p,H),$$

lying on opposite sides of H. These points coincide with control points of a refined control polygon $\tilde{c}$ having the parameters t_k as knots of multiplicity n. Hence, it remains to show that knot insertion does not increase the variation since this implies

$$V(p,H) \leq V(\tilde{c},H) \leq V(c,H).$$

Without going into details, knot insertion repeats the following two steps (not necessarily in any particular order):

(i) placing an additional control point on an edge of the control polygon (splitting edges);

(ii) deleting a control point (cutting corners).

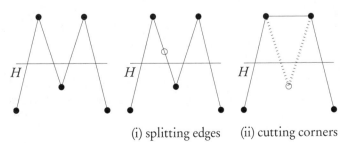

(i) splitting edges (ii) cutting corners

For example, when inserting an interior knot for a cubic spline curve, we place three points on adjacent edges of the control polygon and then delete two of the intermediate vertices. Obviously, neither of the two steps illustrated in the figure increases the variation.

□ **Example:**
Variation diminution can be used to locate the intersections of a spline curve p with a hyperplane H. To this end, we represent p as a union of Bézier segments, recalling that Bernstein polynomials are particular B-splines (cf. Section 4.1). After this preprocessing step, p is represented by a list of polynomial segments p^i, each described by its Bézier control polygon c^i. We now employ an elementary bisection strategy based on the variations

$$V^i = V(c^i, H),$$

which are upper bounds for variations of the Bézier segments.

If V^i is zero, we check whether accidentally one or both of the endpoints of p^i are points of intersection. Since $V(p^i, H) = 0$, intersections in the interior of the parameter interval for p^i are not possible, and we delete p^i from the list.

If $V^i \geq 1$ and the diameter of a bounding box for c^i is $>$ tol, we subdivide p^i at the midpoint of its parameter interval and append both parts in place of p^i to the list.

Upon termination of the subdivision procedure, there are two possibilities for the remaining Bézier segments p^i.

(i) If $V^i = 1$, then the endpoints of c^i, which coincide with the endpoints of p^i, must lie on opposite sides of H. Hence, an intersection of p^i with H exists. Since $V(p^i, H) \leq V^i$, this intersection is unique. Its location is determined by the bounding box for c^i with accuracy tol.

(ii) If $V^i > 1$, the existence of an intersection can be neither guaranteed nor excluded. A more subtle test is needed. This is not surprising since the case of more than one alternation includes the possibility of points of tangency, a numerically critical situation. Fortunately, for small tolerances, Bézier segments with $V^i > 1$ and small bounding boxes are relatively rare.

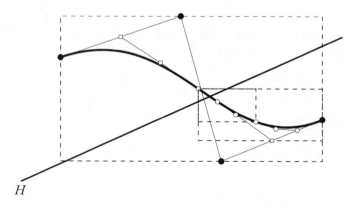

The figure illustrates the subdivision strategy for a planar cubic Bézier segment. Two subdivision steps are performed. We see that the bounding boxes provide an increasingly better approximation to the intersection. When subdivision is terminated, a Newton iteration can be started to determine the exact location. □

6.4 ▪ Algorithms

The algorithms for spline functions easily generalize to curves. We begin by considering the two principal operations.

Evaluation and Differentiation

A point

$$p(s) = \sum_{k=0}^{m-1} c_k\, b_k(s)$$

on a spline curve with knot sequence $\tau : \tau_0, \ldots, \tau_{m+n}$ can be computed by repeatedly inserting s as a new knot until its multiplicity becomes n:

$$\tilde{\tau}_\ell < \tilde{\tau}_{\ell+1} = \cdots = \tilde{\tau}_{\ell+n} = s < \tilde{\tau}_{\ell+n+1} \implies p(s) = \tilde{c}_\ell,$$

where $\tilde{\tau}_\ell$ and $\tilde{c}_k$ denote the modified knots and control points, respectively.

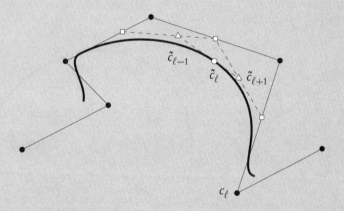

The refined control polygon $\tilde{c}$ is tangent to the curve:

$$p'(s^-) = \frac{n(\tilde{c}_\ell - \tilde{c}_{\ell-1})}{s - \tilde{\tau}_\ell}, \quad p'(s^+) = \frac{n(\tilde{c}_{\ell+1} - \tilde{c}_\ell)}{\tilde{\tau}_{\ell+n+1} - s},$$

where the one-sided derivatives coincide if s is not a knot with multiplicity n of the original knot sequence τ (i.e., if at least one knot is inserted). In this case,

$$p'(s) = \frac{n}{\tilde{\tau}_{\ell+n+1} - \tilde{\tau}_\ell}\left(\tilde{c}_{\ell+1} - \tilde{c}_{\ell-1}\right)$$

is an alternative formula for the tangent vector.

If $s \in (\tilde{\tau}_\ell, \tilde{\tau}_{\ell+n+1})$ is a knot with multiplicity n of $\tilde{\tau}$, the B-splines b_k with $k < \ell$ vanish for parameters $t \geq s$. Hence, $p(s)$ can be regarded as a left endpoint of the spline curve with control points $\tilde{c}_\ell, \tilde{c}_{\ell+1}, \ldots$. The formulas for $p(s)$ and $p'(s^+)$ are thus a direct consequence of endpoint interpolation. Similarly, the formula for $p'(s^-)$ follows by regarding $p(s)$ as a right endpoint of the spline curve with control points $\ldots, \tilde{c}_{\ell-1}, \tilde{c}_\ell$.

If the parametrization is continuously differentiable, $p'(s^-) = p'(s^+)$. Solving this equation for the control point $\tilde{c}_\ell$ and substituting the resulting expression in either of the formulas for the one-sided derivatives yields the formula for $p'(s)$ in terms of $\tilde{c}_{\ell-1}$ and $\tilde{c}_{\ell+1}$.

□ **Example:**

We consider the cubic spline curve p with control points

$$C = \begin{pmatrix} 10 & 4 & 0 & 20 & 20 \\ 10 & 4 & 20 & 20 & 0 \end{pmatrix}^t,$$

knots

$$\tau_0 = 0, 1, 1, 1, 2, 5, 6, 7, 8 = \tau_8,$$

and parameter interval $D_\tau^3 = [1,5]$.

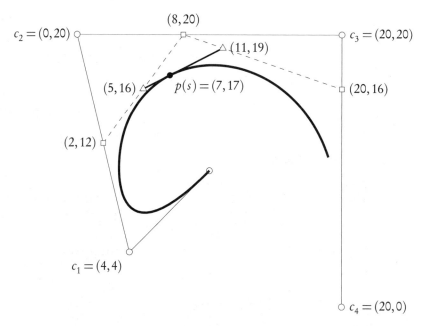

As is illustrated in the figure, we determine the point $p(s)$ and the tangent vector $p'(s)$ for $s = 3 \in [2,5]$. To this end, we insert s three times as new knot. Starting with the relevant control points $c_1, \ldots, c_4$ for the knot interval $[2,5]$, the new control points are generated with a triangular scheme. After having inserted s twice, we can compute

$$p'(3) = \frac{3}{5-2}((11,19) - (5,16)) = (6,3).$$

The point on the curve is then obtained by forming one last convex combination:

$$p(3) = \tilde{c}_4 = \frac{3-2}{5-2}(11,19) + \frac{5-3}{5-2}(5,16) = (7,17),$$

where the weights correspond to the relative position of s in the knot interval $[2,5]$. □

6.4. Algorithms

It is often possible to process the polynomial segments of a spline representation simultaneously. This can be done most efficiently by converting to Bézier form, using repeated knot insertion.

Bézier Form

The Bézier form of a spline curve parametrized by $p = \sum_{k=0}^{m-1} c_k b_k \in S_\tau^n$ is obtained by raising the multiplicity of each knot τ_k in the parameter interval $D_\tau^n = [\tau_n, \tau_m]$ to n. Then, for t in a nondegenerate parameter interval $[\tilde{\tau}_\ell, \tilde{\tau}_{\ell+1}] \subseteq D_{\tilde{\tau}}^n$ of the refined knot sequence $\tilde{\tau}$,

$$p(t) = \sum_{k=0}^{n} \tilde{c}_{\ell-n+k}\, b_k^n(s), \quad s = \frac{t - \tilde{\tau}_\ell}{\tilde{\tau}_{\ell+1} - \tilde{\tau}_\ell} \in [0,1],$$

where b_k^n are the Bernstein polynomials and $\tilde{c}_k$ the control points with respect to $\tilde{\tau}$. Hence, up to linear reparametrization (which is immaterial for the shape of the curve), the spline segments have Bézier form.

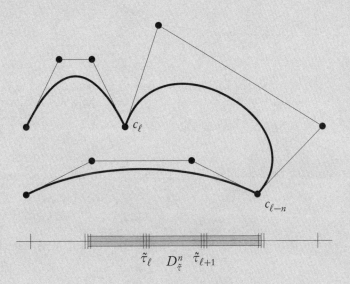

As shown in the figure, every nth control point lies on the curve separating the Bézier segments. Thus, by converting to Bézier form, we can apply polynomial algorithms simultaneously on the different knot intervals.

We already remarked in Section 4.1 that Bernstein polynomials are particular B-splines. If $\tilde{\tau}$ has n-fold knots at both endpoints of a knot interval $D_\ell = [\tilde{\tau}_\ell, \tilde{\tau}_{\ell+1}]$, i.e., if

$$\tilde{\tau}_{\ell-n} < \tilde{\tau}_{\ell-n+1} = \cdots = \tilde{\tau}_\ell < \tilde{\tau}_{\ell+1} = \cdots = \tilde{\tau}_{\ell+n} < \tilde{\tau}_{\ell+n+1},$$

then the relevant B-splines $\tilde{b}_k$ for D_ℓ ($k = \ell - n, \ldots, n$) coincide with the transformed Bernstein polynomials on D_ℓ. As a consequence of the B-spline recursion,

$$\tilde{b}_{\ell-n}(t) = \left(\frac{\tilde{\tau}_{\ell+1} - t}{\tilde{\tau}_{\ell+1} - \tilde{\tau}_\ell}\right)^n = b_0^n(s), \quad \tilde{b}_\ell(t) = \left(\frac{t - \tilde{\tau}_\ell}{\tilde{\tau}_{\ell+1} - \tilde{\tau}_\ell}\right)^n = b_n^n(s)$$

for $t \in D_\ell \Leftrightarrow s \in [0,1]$. We note that these two B-splines have support also on the neighboring intervals, thus ensuring continuity of the curve at the junction points. For $0 < i < n$, the B-spline $\tilde{b}_{\ell-n+i}$ has an $(n+1-i)$-fold knot at τ_ℓ and an $(i+1)$-fold knot at $\tau_{\ell+1}$ and, thus, $\tilde{b}_{\ell-n+i}(t) = b_i^n(s)$; those B-splines are supported entirely on D_ℓ.

☐ **Example:**

As an example, we consider the closed quadratic spline curve with control points and extended knot sequence shown in the figure.

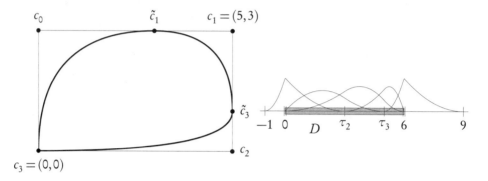

To convert to Bézier form, we double the knots $\tau_2 = 3$ and $\tau_3 = 5$, taking into account the periodicity of the extended partition and the control points:

$$\tau_{k+4} = \tau_k + 6, \quad c_{k+4} = c_k.$$

Applying the knot insertion algorithm first for $s = \tau_2$, only the index $k = 1$ satisfies

$$\tau_k < s = 3 < \tau_{k+2}.$$

Hence, inserting $s = 3$ yields only one new control point, namely,

$$\tilde{c}_1 = \frac{3}{5}(5,3) + \frac{2}{5}(0,3) = (3,3).$$

Similarly, inserting $s = 5$, we compute the control point

$$\tilde{c}_3 = \frac{2}{3}(5,0) + \frac{1}{3}(5,3) = (5,1).$$

and obtain the Bézier control polygon

$$\tilde{c}_0 = c_0, \tilde{c}_1, \tilde{c}_2 = c_1, \tilde{c}_3, \tilde{c}_4 = c_2, \tilde{c}_5 = c_3.$$

We notice that the control points on the curve, which separate the spline segments, lie on the initial control polygon which touches the curve at these points. This is a particular feature of quadratic spline curves. The conversion is very simple in this case. The Bézier control points separating the curve segments at a simple knot τ_ℓ are computed by dividing the segment $[c_{\ell-2}, c_{\ell-1}]$ in the ratio

$$(\tau_\ell - \tau_{\ell-1}) : (\tau_{\ell+1} - \tau_\ell).$$

At a double knot $\tau_\ell = \tau_{\ell+1}$, no computations are necessary; $c_{\ell-1}$ is already a separating Bézier control point. □

As a further illustration, we consider the conversion for uniform knots. In this special case, the knot insertion procedure has an intriguing description via a combinatoric labeling scheme.

□ **Example:**
For spline curves parametrized by $p = \sum_k c_k b_k \in S_\tau^n$ with uniform knots τ_k the conversion to Bézier form has a beautiful geometric interpretation due to Ramshaw and is referred to as blossoming. It is illustrated in the figure below for a parametrization of degree $n = 4$.

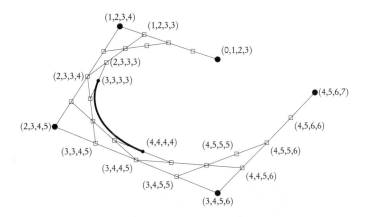

We label the control points by the indices of the inner knots of the corresponding B-splines. For example, c_3 has the label $(4,5,6,7)$ since b_3 has support $[\xi_3, \xi_8]$. Then, we connect pairs of points with labels

$$(a,b,c,d), \quad (b,c,d,e)$$

which have $n-1 = 3$ indices in common. On the resulting line segment, we place $e-a-1$ equally spaced new points and label them with

$$(a+1,b,c,d), \ldots, (e-1,b,c,d),$$

reordering the indices if necessary. The process is repeated, at each stage utilizing the newly generated points, until all possible connections have been formed. In the final configuration, the Bézier segments consist of all points with labels, which have at most two different indices. In particular, the Bézier endpoints have labels with just one index of multiplicity $n = 4$. □

6.5 • Interpolation

Interpolation methods for functions can be applied to curves in an obvious way. Each component is processed separately. Below, we list the most commonly used schemes.

> **Interpolation**
>
> Points p_k and tangent vectors d_k (if provided) can be interpolated with a spline curve at parameter values t_k, using any of the interpolation methods for spline functions. The univariate schemes are applied separately in each component to determine the components of the parametrization $p = \sum_k c_k\, b_k$. Standard choices are cubic Hermite interpolation and cubic spline interpolation with not-a-knot, natural, or clamped boundary conditions.
>
>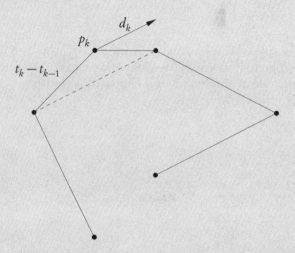
>
> If only points are given, knots τ_j, parameter values t_k, and tangent vectors d_k (if required) have to be determined based on the available information. Basic choices are
>
> - $t_k - t_{k-1} = |p_k - p_{k-1}|$;
> - $t_k = \tau_{k+\ell}$ with the shift ℓ depending on the labeling of the knots;
> - $d_k = (p_{k+1} - p_{k-1})/(t_{k+1} - t_{k-1})$.
>
> More accurate derivative approximations employ local polynomial interpolation. The resulting formulas are used in particular at the endpoints of the parameter interval, where one-sided approximations are needed.

As a first example we consider cubic Hermite interpolation. While the interpolant is not curvature continuous, it can be constructed with a few simple operations without having to solve a linear system.

☐ **Example:**

If tangent vectors d_j are given at all points p_j, we can interpolate each pair of consecutive data with a cubic Bézier segment. By construction, these segments form a spline curve in Bézier form. In particular, two adjacent segments share a common control point lying on the curve.

6.5. Interpolation

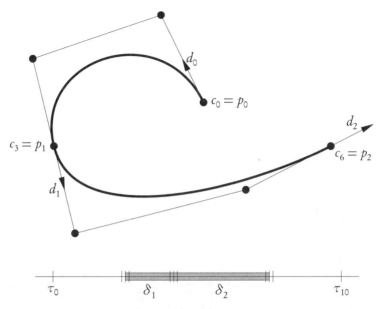

Considering, e.g., the interpolation by the segment interpolating p_0 and p_1, we choose the length of the parameter interval as $\tau_4 - \tau_3 = |p_1 - p_0| = \delta_1$. This means that the parametrization has the form

$$p(t) = \sum_{k=0}^{3} c_k b_k^3(s), \quad s = \frac{t - \tau_3}{\tau_4 - \tau_3},$$

where b_k^3 are the cubic Bernstein polynomials. Accordingly, the interpolation conditions imply

$$p_0 = c_0, \quad p_1 = c_3,$$

and

$$d_0 = \frac{3}{\delta_1}(c_1 - c_0), \quad d_1 = \frac{3}{\delta_1}(c_3 - c_2).$$

Solving the last equations for the two middle control points finally yields

$$c_1 = p_0 + \frac{\delta_1}{3} d_0, \quad c_2 = p_1 - \frac{\delta_1}{3} d_1. \qquad \square$$

Very popular interpolation schemes, also for curves, are natural spline interpolation and its variants. The implementation of these methods is discussed in the following example.

□ **Example:**

To interpolate points $p_0, \ldots, p_M$ by a natural cubic spline curve, we first make a standard selection of the parameter values. We choose t_k so that

$$t_{k+1} - t_k = |p_{k+1} - p_k|, \quad k = 0, \ldots, M-1.$$

In $D = [t_0, t_M]$, the knots $\tau_k = t_{k-3}$, $k = 3, \ldots, m$, $m = M + 3$, are used. Outside of D, we choose three equally spaced knots on each side, keeping the distance of the first and last knot interval. We can now apply the univariate scheme with the boundary conditions $p''(t_0) = p''(t_M) = 0$.

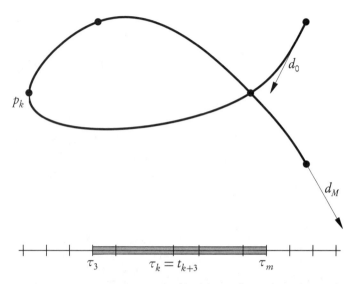

A more accurate approximation is obtained by using either clamped or not-a-knot boundary conditions. In the first case, derivatives are prescribed at the endpoints:

$$p'(t_0) = d_0, \quad p'(t_M) = d_M.$$

The not-a-knot conditions mean that the knots $\tau_4 = t_1$ and $\tau_{m-1} = t_{M-1}$ are removed. As a result, we interpolate using the B-splines $\tilde{b}_0, \ldots, \tilde{b}_M$ corresponding to the reduced knot sequence

$$\tilde{\tau}: \tau_0 < \cdots < \tau_3 = t_0 < \tau_5 < \tau_6 < \cdots < \tau_{m-3} < \tau_{m-2} < t_M = \tau_m < \cdots < \tau_{m+3}$$

Hence, the dimension $\tilde{m} = M + 1$ of the spline space $S^3_{\tilde{\tau}}$ matches the number of interpolation conditions and, as implied by the Schoenberg–Whitney condition, the interpolation problem is well posed. □

Chapter 7

Multivariate Splines

There exist several different concepts for defining multivariate splines. A straightforward approach, as adopted in this book, uses the tensor product formalism. Because of its simplicity, this method is particularly suited for applications, for which computational efficiency is important. All advantages of the univariate B-spline calculus persist in the multivariate setting.

After briefly reviewing basic facts about multivariate polynomials in Sections 7.1 and 7.2, we define multivariate tensor product B-splines and splines in Section 7.3. In Section 7.4, we describe the basic algorithms for evaluation and differentiation. Section 7.5 is devoted to the generalization of univariate approximation schemes. Finally, in Section 7.6, we discuss hierarchical B-spline bases, which are fundamental for adaptive techniques.

7.1 ▪ Polynomials

For a multivariate polynomial, such as

$$x \mapsto p(x) = 3x_1^9 x_2^4 - 6x_1^7 x_2^5,$$

there are essentially two notions of degree. Either one computes the maximum of the sums of the exponents for each monomial,

$$\max(9+4, 7+5) = 13 \quad \text{(total degree)},$$

or one takes the maximum of the vectors of exponents,

$$\max((9,4), (7,5)) = (9,5) \quad \text{(coordinate degree)}.$$

The latter definition is best suited for the tensor product formalism employed in this section.

Multivariate Polynomials

A (real) polynomial p of d variables and coordinate degree $n = (n_1, \ldots, n_d)$ is a linear combination of monomials,

$$p(x) = \sum_{k \leq n} c_k x^k, \quad x^k = x_1^{k_1} \cdots x_d^{k_d},$$

133

with coefficients $c_k \in \mathbb{R}$ and $c_n \neq 0$. The sum is taken over all multi-indices k (nonnegative integer vectors) with $k_\nu \leq n_\nu$; i.e., p is a univariate polynomial of degree n_ν in each coordinate direction.

The d-variate polynomials of coordinate degree $\leq n$ form a linear vector space, denoted by $\mathbb{P}^n$, of dimension $(n_1 + 1) \cdots (n_d + 1)$. More precisely, we write $\mathbb{P}^n(D)$ if the variable x is restricted to a particular domain $D \subseteq \mathbb{R}^d$.

As a first example, we consider the simple multilinear case, i.e., degree ≤ 1 in each coordinate direction.

☐ **Example:**
A bilinear polynomial,
$$p(x_1, x_2) = c_{0,0} + c_{1,0} x_1 + c_{0,1} x_2 + c_{1,1} x_1 x_2,$$
is determined by four coefficients c_k, which match the partial derivatives $\partial_1^{k_1} \partial_2^{k_2} p(0,0)$. Alternatively, p can be specified by its values p_k at the corners
$$k \in K = \{0, 1\} \times \{0, 1\}$$
of the unit square $D = [0, 1] \times [0, 1]$.

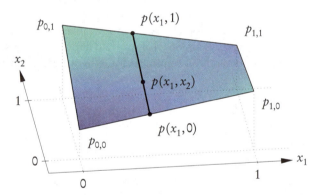

As is illustrated in the figure, we can evaluate p via repeated linear interpolation. Since p is linear along the horizontal boundaries of D,
$$p(x_1, 0) = p_{0,0}(1 - x_1) + p_{1,0} x_1, \quad p(x_1, 1) = p_{0,1}(1 - x_1) + p_{1,1} x_1.$$
Interpolating once more along the vertical segment $[(x_1, 0), (x_1, 1)]$, where p is also linear,
$$p(x_1, x_2) = p(x_1, 0)(1 - x_2) + p(x_1, 1) x_2$$
yields the value of p at an arbitrary point $x \in D$.

The interpolation scheme generalizes to d variables. Carrying out the successive interpolations yields the Lagrange form
$$p(x) = \sum_{k_1=0}^{1} \cdots \sum_{k_d=0}^{1} p_k \prod_{\nu=1}^{d} (1 - x_\nu)^{1 - k_\nu} x_\nu^{k_\nu}.$$

Since the product equals one if $x = k$, and zero if x is another vertex of $[0, 1]^d$, the correctness of this explicit expression for p is easily confirmed. ☐

For a univariate polynomial of degree $\leq n$, the $(n+1)$th derivative vanishes. Multivariate polynomials of coordinate degree $\leq (n_1, \dots, n_d)$ can be characterized in an analogous fashion:

$$p \in \mathbb{P}^n \quad \Leftrightarrow \quad \partial_\nu^{n_\nu+1} p = 0 \; \forall \, \nu = 1, \dots, d \,,$$

where ∂_ν denotes the partial derivative with respect to the νth variable.

To prove this statement, we note that for

$$p(x) = \sum_k c_k x^k$$

the $(n_\nu+1)$th partial derivative with respect to x_ν annihilates all monomials x^k, for which the exponent of x_ν is $\leq n_\nu$. Therefore,

$$\partial_\nu^{n_\nu+1} p = \sum_{k_1} \cdots \sum_{k_\nu > n_\nu} \cdots \sum_{k_d} c_k \left(k_\nu \cdots (k_\nu - n_\nu) \right) x^{(k_1, \dots, k_\nu - n_\nu - 1, \dots, k_d)} \,.$$

As a consequence,

$$\partial_\nu^{n_\nu+1} p = 0 \; \forall \, \nu \quad \Leftrightarrow \quad c_k = 0 \text{ if } k_\nu > n_\nu \text{ for some } \nu \,.$$

Only linear combinations of the monomials with exponents $k_\nu \leq n_\nu$ for all ν are in the common nullspace of the partial derivatives $\partial_\nu^{n_\nu+1}$.

As for the univariate theory, Bernstein polynomials play an important role because of their favorable numerical and geometric properties. The univariate definition generalizes easily.

> **Multivariate Bernstein Polynomials**
>
> The d-variate Bernstein polynomials of coordinate degree $(n_1, \dots, n_d)$ are products of univariate Bernstein polynomials:
>
> $$b_k^n(x) = \prod_{\nu=1}^d b_{k_\nu}^{n_\nu}(x_\nu), \quad 0 \leq k_\nu \leq n_\nu,$$
>
> for x in the standard parameter domain $D = [0,1]^d$. They form a basis for $\mathbb{P}^n(D)$, which is symmetric with respect to the domain boundaries.
>
> The properties of univariate Bernstein polynomials extend to d variables. In particular,
>
> $$b_k^n \geq 0, \quad \sum_k b_k^n = 1.$$
>
> Moreover, at a vertex $\sigma \in \{0,1\}^d$ of D,
>
> $$b_{(\sigma_1 n_1, \dots, \sigma_d n_d)}^n(\sigma) = 1,$$
>
> and all other Bernstein polynomials vanish at σ.

The following example illustrates the properties of the Bernstein basis.

☐ **Example:**
The figure shows the nine biquadratic Bernstein polynomials $b_k^{(2,2)}$, $0 \leq k_1, k_2 \leq 2$.

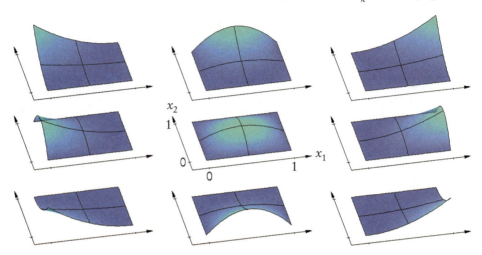

In particular, we notice the endpoint interpolation property. To each corner of the square $D = [0,1]^2$, there corresponds one biquadratic Bernstein polynomial, which has the value 1 there. For example, $b_{(2,0)}^{(2,2)}$ (lower right) equals 1 at $(x_1, x_2) = (1,0)$. ☐

7.2 • Polynomial Approximation

As a basic prerequisite for studying approximation with multivariate splines, we analyze the error of local polynomial approximations. Since the polynomials of coordinate degree $\leq (n_1, \ldots, n_d)$ form the common kernel of the pure partial derivatives of order $n_\nu + 1$, one expects a bound in terms of the norms of these derivatives. Surprisingly, multivariate Taylor polynomials do not yield an estimate of this type, as was illustrated by Reif with the following example.

☐ **Example:**
For $f(x) = (\varepsilon + x_1 + x_2)^{7/2}$ with $\varepsilon \in (0,1]$, the Taylor polynomial of degree $\leq n = (2,2)$ at the origin equals

$$p(x) = \varepsilon^{7/2} + \frac{7}{2}\varepsilon^{5/2}(x_1+x_2) + \frac{35}{8}\varepsilon^{3/2}(x_1^2+2x_1x_2+x_2^2) + \frac{105}{16}\varepsilon^{1/2}(x_1^2 x_2 + x_1 x_2^2) + \frac{105}{64}\varepsilon^{-1/2}x_1^2 x_2^2.$$

Hence, because of the last term,

$$\max_{x \in [0,1]^2} |f(x) - p(x)| \geq c\,\varepsilon^{-1/2}$$

for $\varepsilon \to 0$. On the other hand, the partial derivatives

$$\partial^{(3,0)} f(x) = \partial^{(0,3)} f(x) = \frac{105}{8}\varepsilon^{1/2}(\varepsilon + x_1 + x_2)^{1/2}$$

are bounded on $[0,1]^2$, uniformly in ε. This shows that the Taylor remainder cannot be estimated in terms of the derivatives characterizing $\mathbb{P}^n$. ☐

7.2. Polynomial Approximation

Approximations which yield the natural error bound are provided by orthogonal polynomials.

Error of Polynomial Approximation

For a hyperrectangle $R = [a_1, b_1] \times \cdots \times [a_d, b_d]$, the error of the orthogonal projection $P^n f \in \mathbb{P}^n(R)$ defined by

$$\int_R f q = \int_R (P^n f) q \quad \forall q \in \mathbb{P}^n(R)$$

can be estimated by

$$|f(x) - (P^n f)(x)| \le c(d, n) \sum_{\nu=1}^{d} h_\nu^{n_\nu + 1} \|\partial_\nu^{n_\nu + 1} f\|_{\infty, R} \quad \forall x \in R,$$

where h_ν denotes the width of R in the νth direction.

The explicit bound and its elegant proof is due to Reif. The arguments are divided into several steps.

(i) First, we note that the estimate is invariant under translation and scaling. Hence, we may assume that $R = [0, 1]^d$.

(ii) Considering the univariate case already yields the key inequalities. With $p_0, \ldots, p_n$ an orthonormal basis for $\mathbb{P}^n([0, 1])$, we have

$$P^n f = \sum_{k=0}^{n} \left(\int_0^1 f\, p_k \right) p_k$$

by definition of an orthogonal projection. It follows that

$$\|P^n f\|_{\infty, [0,1]} \le c_1(n) \|f\|_{\infty, [0,1]}$$

with $c_1(n) = \sum_k \left(\int |p_k| \right) \|p_k\|_{\infty, [0,1]}$. The boundedness of the univariate projector P^n yields the desired bound for the error by a standard argument:

$$\|f - P^n f\|_{\infty, [0,1]} \le \|f - p\|_{\infty, [0,1]} + \|P^n(p - f)\|_{\infty, [0,1]} \le (1 + c_1(n)) \|f - p\|_{\infty, [0,1]}$$

with p any polynomial of degree $\le n$. Choosing for p the Taylor polynomial, the right side is

$$\le c_2(n) \|f^{(n+1)}\|_{\infty, [0,1]}, \quad c_2(n) = \frac{1 + c_1(n)}{(n+1)!}$$

in view of the bound for the Taylor remainder.

(iii) In order to generalize the argument, we now derive a convenient representation for the multivariate projection. We define

$$(P_\nu f)(x_1, \ldots, x_d) = \sum_{i=0}^{n_\nu} \left(\int_0^1 f(x_1, \ldots, x_{\nu-1}, t_\nu, x_{\nu+1}, \ldots, x_d) p_i(t_\nu)\, dt_\nu \right) p_i(x_\nu), \quad \nu = 1, \ldots, d,$$

i.e., P_ν is the univariate projector applied to the νth variable with the other variables held fixed. Clearly, the univariate estimates remain valid for $P_\nu f$ and $f - P_\nu f$ with $[0,1]$ replaced by $[0,1]^d$ and $f^{(n+1)}$ by $\partial_\nu^{n_\nu+1} f$. Moreover, it is easily seen that

$$P^{(n_1,\dots,n_d)} = P_1 \cdots P_d.$$

We just note that the d-variate polynomials $x \mapsto p_k(x) = \prod_{\nu=1}^d p_{k_\nu}(x_\nu)$ form an orthonormal basis and

$$P^n f = \sum_{k_1=0}^{n_1} \cdots \sum_{k_d=0}^{n_d} \left(\int f(t)\, p_{k_1}(t_1) \cdots p_{k_d}(t_d)\, dt_1 \cdots dt_d \right) p_k.$$

Writing the integral as a product of univariate integrals with the aid of Fubini's theorem yields the factorization $P_1 \cdots P_d$ of the multivariate projector P^n. For example, for $d = 2$,

$$(P^n f)(x) = \sum_{k_1=0}^{n_1} \left(\int \left[\sum_{k_2=0}^{n_2} \left(\int f(t_1, t_2) p_{k_2}(t_2)\, dt_2 \right) p_{k_2}(x_2) \right] p_{k_1}(t_1)\, dt_1 \right) p_{k_1}(x_1),$$

where $[\dots] = (P_2 f)(t_1, x_2)$ and thus $P^n f = P_1(P_2 f)$.

(iv) From the product form of P^n, we deduce that

$$
\begin{aligned}
f - P^n f = {} & [f - P_1 f] \\
& + P_1[f - P_2 f] \\
& + \cdots \\
& + P_1 \cdots P_{d-1}[f - P_d f].
\end{aligned}
$$

With this decomposition of the error, the univariate estimates imply

$$|f(x) - (P^n f)(x)| \le \sum_{\nu=1}^d c_1(n_1) \cdots c_1(n_{\nu-1}) \left[c_2(n_\nu) \|\partial_\nu^{n_\nu+1} f\|_{\infty, [0,1]^d} \right]$$

for any $x \in R = [0,1]^d$.

7.3 ▪ Splines

Multivariate B-splines are defined on a tensor product grid determined by knot sequences

$$\xi_\nu : \dots \xi_{\nu,0}, \xi_{\nu,1}, \dots, \quad \nu = 1, \dots, d,$$

in each coordinate direction.

Multivariate B-Splines

The d-variate B-splines of degree $(n_1, \dots, n_d)$ with respect to the knot sequences ξ are products of univariate B-splines:

$$b_{k,\xi}^n(x) = \prod_{\nu=1}^d b_{k_\nu, \xi_\nu}^{n_\nu}(x_\nu).$$

Their knots in the νth coordinate direction are $\xi_{\nu,k_\nu}, \dots, \xi_{\nu,k_\nu+n_\nu+1}$.

7.3. Splines

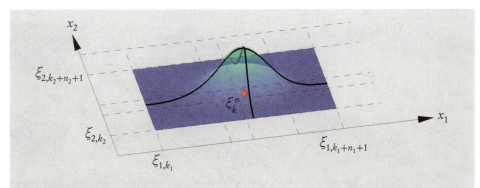

The multivariate knot average

$$\xi_k^n = \left(\xi_{1,k_1}^{n_1}, \ldots, \xi_{d,k_d}^{n_d}\right), \quad \xi_{\nu,k_\nu}^{n_\nu} = \left(\xi_{\nu,k_\nu+1} + \cdots + \xi_{\nu,k_\nu+n_\nu}\right)/n_\nu$$

is often used to identify multivariate B-splines on a grid and can be viewed as weighted center of their support.

The properties of univariate B-splines extend to d variables in a canonical way. The multivariate B-spline $b_{k,\xi}^n$

- vanishes outside the hyperrectangle $[\xi_{1,k_1}, \xi_{1,k_1+n_1+1}) \times \cdots \times [\xi_{d,k_d}, \xi_{d,k_d+n_d+1})$;

- is a nonnegative polynomial of coordinate degree $\leq n$ on each grid cell $[\xi_{1,\ell_1}, \xi_{1,\ell_1+1}) \times \cdots \times [\xi_{d,\ell_d}, \xi_{d,\ell_d+1})$;

- coincides with a multiple of $b_{k_\nu,\xi_\nu}^{n_\nu}$ on each line parallel to the νth coordinate axis.

For multivariate B-splines, nonuniform knots do not provide the same flexibility as in one variable. A refinement of one of the knot sequences ξ_ν has a global effect. Hence, a tensor product grid does not permit truly local modifications. However, to overcome this disadvantage, nested grids can be used as described in Section 7.6. In particular in this context, uniform B-splines play an important role. They are scaled translates of the standard uniform B-spline b^n, which has knots $0, \ldots, n_\nu + 1$ in the ν-th coordinate direction:

$$b_{k,\xi}^n(x) = b^n((x - (\xi_{1,k_1}, \ldots, \xi_{d,k_d}))/h) = \prod_{\nu=1}^d b^{n_\nu}((x_\nu - \xi_{\nu,k_\nu})/h),$$

where b^{n_ν} are the univariate standard uniform B-splines with support $[0, n_\nu + 1]$ and ξ_ν, $\nu = 1, \ldots, d$, are knot sequences with grid width h. This definition is illustrated in the following example.

☐ **Example:**
The figure shows B-splines with coordinate degrees ≤ 2 and uniform knots $\xi_{\nu,k} = kh$.

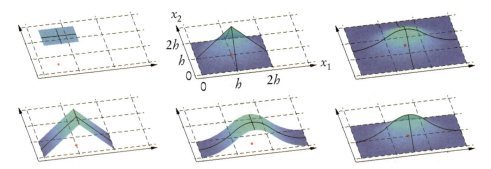

The examples in the top row have equal coordinate degrees. B-splines with $n_1 \neq n_2$, as shown in the bottom row, are rarely used. They do arise, however, when taking partial derivatives. □

As in one variable, the standard uniform B-spline b^n can be defined by an averaging process. Starting with the characteristic function $b^{(0,\ldots,0)}$ of $[0,1]^d$, we integrate successively in the directions of the unit vectors $u = (1,0,\ldots,0),\ldots,(0,\ldots,0,1)$,

$$b^{(n+u)}(x) = \int_0^1 b^n(x - tu)\,dt,$$

repeating the average in the νth coordinate direction n_ν times. This definition has an obvious generalization. We average in arbitrary directions $(u_1,\ldots,u_m)$. The figure gives an example of the resulting so-called box-splines introduced by de Boor and DeVore.

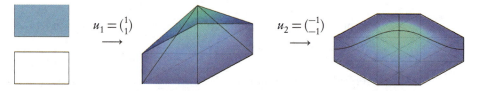

We see that the partition consists no longer just of hypercubes as for uniform B-splines. It is generated by the directions used in the averaging process. If many different directions u_ℓ are used, the pattern of discontinuities becomes complicated. From a practical point of view, this is a slight drawback. The mathematical theory, however, is truly beautiful.

The definition of multivariate splines is analogous to the univariate case. Again, we limit the admissible knot sequences by insisting on n-regularity. In accordance with the tensor product formalism, this means that each of the knot sequences ξ_ν, $\nu = 1,\ldots,d$, is n_ν-regular. Moreover,

$$D_\xi^n = D_{\xi_1}^{n_1} \times \cdots \times D_{\xi_d}^{n_d}$$

is the parameter hyperrectangle. With these notations, we define splines in the usual way.

Multivariate Splines

A multivariate spline p of degree $\leq n = (n_1,\ldots,n_d)$ is a linear combination of the B-splines corresponding to n-regular knot sequences $\xi = (\xi_1,\ldots,\xi_d)$:

$$p(x) = \sum_k c_k b_{k,\xi}^n(x), \quad x \in D_\xi^n.$$

7.3. Splines

> The coefficients are unique; i.e., the B-splines b_k restricted to the parameter hyperrectangle D_ξ^n form a basis for the spline space denoted by S_ξ^n.
>
> Spline functions on a subdomain $D \subset D_\xi^n$ are obtained simply by restricting the variable x to the smaller set; the corresponding spline space is denoted by $S_\xi^n(D)$. A basis consists of the relevant B-splines b_k, $k \sim D$, which have some support in D.

For implementing algorithms, it is not convenient to keep track of the relevant B-splines by storing lists of indices. Instead, for finite knot sequences

$$\xi_\nu : \xi_{\nu,0},\ldots,\xi_{\nu,m_\nu+n_\nu},$$

we represent $p \in S_\xi^n(D)$ with $D \subseteq D_\xi^n$ by a linear combination

$$\sum_{k_1=0}^{m_1-1} \cdots \sum_{k_d=0}^{m_d-1} c_k b_k$$

and set coefficients c_k of irrelevant B-splines ($k \not\sim D$) to zero. In this way, we can work entirely with rectangular coefficient arrays. The figure illustrates this procedure for a bivariate domain and coordinate degree $n = (2,2)$. Dots are placed at the knot averages of the relevant biquadratic B-splines, circles indicate irrelevant B-splines. The boundary of the hyperrectangle $D_\xi^n \supset D$ is shown with a solid line.

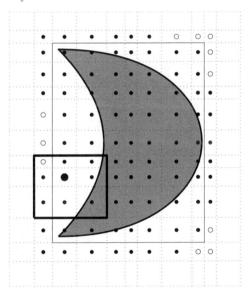

The spline space $S_\xi^n(D)$ contains all multivariate polynomials of coordinate degree $\leq n$. To prove this, we use Marsden's identity in each coordinate direction:

$$(x_\nu - y_\nu)^{n_\nu} = \sum_{k_\nu \in \mathbb{Z}} \psi_{k_\nu,\xi_\nu}^{n_\nu}(y_\nu) b_{k_\nu,\xi_\nu}^{n_\nu}(x_\nu), \quad \psi_{k_\nu,\xi_\nu}^{n_\nu}(y_\nu) = (\xi_{\nu,k_\nu+1} - y_\nu) \cdots (\xi_{\nu,k_\nu+n_\nu} - y_\nu),$$

where $\nu = 1,\ldots,d$, and we have extended the knot sequences so that $\lim_{\ell \to \pm\infty} \xi_{\nu,\ell} = \pm\infty$. Multiplying these identities yields a multivariate version of Marsden's formula:

$$(x-y)^n = \sum_{k_1} \cdots \sum_{k_d} \left(\prod_\nu \psi_{k_\nu,\xi_\nu}^{n_\nu}(y_\nu) \right) b_{k,\xi}^n(x).$$

Of course, if the variable x is restricted to D, the sum has to be taken only over the relevant indices $k \sim D$, proving that $\mathbb{P}^n(D) \subseteq S_\xi^n(D)$ (equality holds if D is contained in a single grid cell).

As a consequence, the relevant B-splines $b_{k,\xi}^n$, $k \sim D$, are linearly independent on any open subset of D. The argument is completely analogous to the univariate case. For any grid cell D_ℓ, which intersects the interior of D, exactly $s = \prod_\nu (n_\nu + 1)$ B-splines are relevant for $D_\ell \cap D$. Since $s = \dim \mathbb{P}^n$ and all multivariate polynomials of coordinate degree $\leq n$ can be represented as linear combinations of these B-splines on $D_\ell \cap D$, these B-splines are linearly independent.

The linear independence on an arbitrary open subset U of D follows easily. We assume that a linear combination

$$\sum_{k \sim U} c_k \, b_{k,\xi}^n$$

of the relevant B-splines for U vanishes on U. By considering the intersections $D_\ell \cap U$ with grid cells D_ℓ, we conclude that all coefficients must be zero.

7.4 ▪ Algorithms

The basic univariate algorithms are easily extended with the aid of the tensor product formalism. In effect, one applies the corresponding univariate schemes successively in each coordinate direction.

Evaluation

A d-variate spline $p = \sum_{k \sim D} c_k \, b_k \in S_\xi^n(D)$ can be evaluated at a point $x \in D$ with $\xi_{\nu,\ell_\nu} \leq x_\nu < \xi_{\nu,\ell_\nu+1}$ by applying the de Boor algorithm in each variable. For $\nu = 1, \dots, d$, we modify the coefficients

$$a_{k_\nu} = c_{(\ell_1, \dots, \ell_{\nu-1}, k_\nu, k_{\nu+1}, \dots, k_d)}, \quad \ell_\nu - n_\nu \leq k_\nu \leq \ell_\nu,$$

simultaneously for all $k_\mu = \ell_\mu - n_\mu, \dots, \ell_\mu$, $\mu > \nu$. We compute, for $i = n_\nu, \dots, 1$,

$$a_{\ell_\nu - j} \leftarrow \gamma \, a_{\ell_\nu - j} + (1 - \gamma) a_{\ell_\nu - j - 1}, \qquad \gamma = \frac{x_\nu - \xi_{\nu, \ell_\nu - j}}{\xi_{\nu, \ell_\nu - j + i} - \xi_{\nu, \ell_\nu - j}} \quad \text{for } j = 0, \dots, i-1,$$

according to the triangular univariate evaluation scheme. This leads to the modified coefficients $c_{(\ell_1, \dots, \ell_\nu, k_{\nu+1}, \dots, k_d)}$. The final computed coefficient c_ℓ then equals $p(x)$.

Considering only the relevant indices for the point x, we write

$$p(x) = \sum_{k_2, \dots, k_d} \left[\sum_{k_1} c_k \, b_{k_1, \xi_1}^{n_1}(x_1) \right] b_{(k_2, \dots, k_d),(\xi_2, \dots, \xi_d)}^{(n_2, \dots, n_d)}(x_2, \dots, x_d), \qquad \cdot$$

where $\ell_\mu - n_\mu \leq k_\mu \leq \ell_\mu$. We fix the indices k_μ with $\mu > 1$ and indicate this by writing $a_{k_1} = c_k$. Then, the sum in brackets is a univariate spline, which can be evaluated with the de Boor algorithm. As a result, we obtain modified coefficients

$$c_{(\ell_1, k_2, \dots, k_d)} = [\dots].$$

7.4. Algorithms

The dimension of the coefficient array to be processed has been reduced. Repeating the process with the remaining variables $x_2, \ldots, x_d$,

$$c_{(\ell_1, k_2, \ldots, k_d)} \xrightarrow{\nu=2} c_{(\ell_1, \ell_2, k_3, \ldots, k_d)} \cdots \xrightarrow{\nu=d} c_{(\ell_1, \ell_2, \ldots, \ell_d)},$$

finally yields the value $p(x)$.

For uniform knot sequences ξ_ν with grid width h, the formulas simplify slightly. Writing $x = \xi_\ell + th$ with $t \in [0,1)^d$ for a point in D,

$$\gamma = \frac{j + t_\nu}{i}, \quad (1 - \gamma) = \frac{i - j - t_\nu}{i},$$

in the evaluation algorithm. Moreover, since the denominators do not change in the innermost loop, we can replace γ and $1 - \gamma$ by $j + t_\nu$ and $i - j - t_\nu$ and divide by $n! = n_1! \cdots n_d!$ at the very end of the algorithm. We illustrate the resulting scheme for evaluating uniform splines with a concrete example.

□ **Example:**

We evaluate the uniform bivariate spline $p = \sum_k c_k\, b_k \in S_\xi^{(2,3)}$ with $\xi_{\nu, k_\nu} = k_\nu$ and coefficients

$$c_{(4,5)} = 8 \quad 1 \quad 1 \quad 0$$
$$\cdots \qquad 0 \quad 5 \quad 9 \quad 4 \qquad \cdots$$
$$8 \quad 9 \quad 9 \quad 8 = c_{(6,8)}$$

at $x = (6.5, 8)$.

In the first evaluation step corresponding to the variable x_1 the de Boor algorithm operates on the columns of the array of relevant coefficients shown above. With $a_{k_1} = c_{(k_1, k_2)}$ (column index k_2 held fixed), $\ell = (6, 8)$, and $t = (1/2, 0)$,

$$\text{step } i = 2: \quad a_6 \leftarrow (1/2)a_6 + (3/2)a_5, \quad a_5 \leftarrow (3/2)a_5 + (1/2)a_4,$$
$$\text{step } i = 1: \quad a_6 \leftarrow (1/2)a_6 + (1/2)a_5,$$

where divisions are postponed. Accordingly, the two steps yield the following modifications of the relevant coefficient array:

$$\begin{matrix} 8 & 1 & 1 & 0 \\ 0 & 5 & 9 & 4 \\ 8 & 9 & 9 & 8 \end{matrix} \rightarrow \begin{matrix} 4 & 8 & 14 & 6 \\ 4 & 12 & 18 & 10 \end{matrix} \rightarrow c_{(6,5)} = 4 \quad 10 \quad 16 \quad 8 = c_{(6,8)}.$$

The second evaluation step computes the cubic spline

$$4b_{5,\xi_2}^3 + 10 b_{6,\xi_2}^3 + 16 b_{7,\xi_2}^3 + 8 b_{8,\xi_2}^3$$

at $x_2 = 8$. Now the triangular de Boor scheme without divisions yields

$$\begin{array}{ccccccc}
4 & & & & & & \\
 & \searrow 1 & & & & & \\
10 & \underline{\quad 2 \quad} & 24 & & & & \\
 & \quad 2 \quad & & \searrow 1 & & & \\
16 & \underline{\quad 1 \quad} & 36 & \underline{\quad 1 \quad} & 60 & & \\
 & \quad 3 \quad & & \quad 2 \quad & & \searrow 1 & \\
8 & \underline{\quad 0 \quad} & 48 & \underline{\quad 0 \quad} & 72 & \underline{\quad 0 \quad} & 60.
\end{array}$$

The final modified coefficient $c_{(6,8)} = 60$ is divided by $(2,3)! = 2 \cdot 6$ to obtain $p(6.5, 8) = 5$.

We notice that the last row is not needed. Since $x_2 = 8$ coincides with a knot, the B-spline b^3_{8,ξ_2} does not contribute to $p(x)$. However, as we remarked earlier, such special cases are usually ignored in implementations. □

The differentiation of multivariate splines is as simple as in the univariate case. We just have to form appropriate differences of the B-spline coefficients.

Differentiation

The partial derivatives ∂_ν, $\nu = 1, \ldots, d$, of a d-variate B-spline of degree $n = (n_1, \ldots, n_d)$ are differences of B-splines of lower degree:

$$\partial_\nu b^n_{k,\xi} = \alpha^{n_\nu}_{k_\nu,\xi_\nu} b^{n-e_\nu}_{k,\xi} - \alpha^{n_\nu}_{k_\nu+1,\xi_\nu} b^{n-e_\nu}_{k+e_\nu,\xi}, \quad \alpha^{n_\nu}_{k_\nu,\xi_\nu} = \frac{n_\nu}{\xi_{\nu,k_\nu+n_\nu} - \xi_{\nu,k_\nu}},$$

with e_ν the νth unit vector.

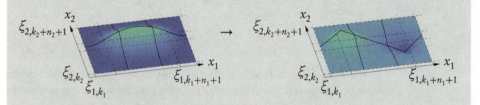

Accordingly, the partial derivative of a multivariate spline in $S^n_\xi(D)$ is given by

$$\partial_\nu \sum_{k \sim D} c_k b^n_{k,\xi} = \sum_{k \sim D} \alpha^{n_\nu}_{k_\nu,\xi_\nu} (c_k - c_{k-e_\nu}) b^{n-e_\nu}_{k,\xi},$$

where on both sides the sum is taken over the relevant B-splines for the domain D.

Applying $\partial_\nu = \partial/\partial x_\nu$ to the product

$$b^n_{k,\xi}(x) = b^{n_1}_{k_1,\xi_1}(x_1) \cdots b^{n_\nu}_{k_\nu,\xi_\nu}(x_\nu) \cdots b^{n_d}_{k_d,\xi_d}(x_d),$$

only the νth factor changes. In view of the univariate formula

$$\frac{d}{dx} b^n_{k,\xi}(x) = \alpha^n_{k,\xi} b^{n-1}_{k,\xi}(x) - \alpha^n_{k+1,\xi} b^{n-1}_{k+1,\xi}(x),$$

the factor becomes

$$\alpha^{n_\nu}_{k_\nu,\xi_\nu} b^{n_\nu-1}_{k_\nu,\xi_\nu}(x_\nu) - \alpha^{n_\nu}_{k_\nu+1,\xi_\nu} b^{n_\nu-1}_{k_\nu+1,\xi_\nu}(x_\nu).$$

Adjoining the other factors and noting that $n - e_\nu = (n_1, \ldots, n_\nu - 1, \ldots, n_d)$ yields the asserted formula for $\partial_\nu b^n_{k,\xi}$.

For the partial derivative with respect to the νth variable of a spline $p = \sum_k c_k b^n_{k,\xi}$, we obtain

$$\partial_\nu p = \sum_{k \sim D} c_k \left(\alpha^{n_\nu}_{k_\nu, \xi_\nu} b^{n-e_\nu}_{k,\xi} - \alpha^{n_\nu}_{k_\nu+1, \xi_\nu} b^{n-e_\nu}_{k+e_\nu, \xi} \right).$$

Substituting $k \leftarrow k - e_\nu$ in the second summand, the difference operation can be shifted from the B-splines to the coefficients.

☐ **Example:**

We apply the gradient and the Laplace operator to a bivariate uniform quadratic B-spline $b^{(2,2)}_{k,\xi}$. Since $\xi_{\nu,\ell+1} = \xi_{\nu,\ell} + h$ for all ℓ, the coefficient α in the differentiation formula equals h^{-1}. Hence, we obtain

$$\operatorname{grad} b^{(2,2)}_{k,\xi} = h^{-1} \left(b^{(1,2)}_{(k_1,k_2),\xi} - b^{(1,2)}_{(k_1+1,k_2),\xi},\, b^{(2,1)}_{(k_1,k_2),\xi} - b^{(2,1)}_{(k_1,k_2+1),\xi} \right).$$

Then, with $\Delta = \operatorname{div}\operatorname{grad} = \partial_1^2 + \partial_2^2$,

$$h^2 \Delta b^{(2,2)}_{k,\xi} = b^{(0,2)}_{(k_1,k_2),\xi} - 2 b^{(0,2)}_{(k_1+1,k_2),\xi} + b^{(0,2)}_{(k_1+2,k_2),\xi} + b^{(2,0)}_{(k_1,k_2),\xi} - 2 b^{(2,0)}_{(k_1,k_2+1),\xi} + b^{(2,0)}_{(k_1,k_2+2),\xi}.$$

We notice in particular that partial differentiation decreases the coordinate degree of the B-spline only in the corresponding direction.

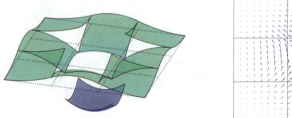 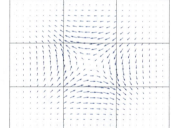

The figure shows $\Delta b^{(2,2)}_{k,\xi}$ (left) and $\operatorname{grad} b^{(2,2)}_{k,\xi}$ visualized as a vector field (right). ☐

7.5 ▪ Approximation Methods

For general domains D, the construction of appropriate approximation schemes can be delicate. The main difficulty is to maintain the optimal accuracy near free-form boundaries. The simple tensor product formalism applies only if D is a hyperrectangle. In this case, the generalization of univariate procedures is straightforward.

Approximation

Univariate spline approximation schemes,

$$f \mapsto p = \sum_{k=0}^{m-1} c_k \, b_k \,,$$

which can be described by a matrix operation $c = Af$, can be combined to construct approximations with multivariate splines

$$p = \sum_{k_1=0}^{m_1-1} \cdots \sum_{k_d=0}^{m_d-1} c_k \, b_{k,\xi}^n \in S_\xi^n \,,$$

where $\xi_\nu : \xi_{\nu,0}, \dots, \xi_{\nu,m_\nu+n_\nu}$. For a d-dimensional array of data f, the multivariate scheme $f \mapsto p$ is defined by

$$c_{(k_1,\dots,k_d)} = \sum_{\ell_1} \cdots \sum_{\ell_d} a_{k_1,\ell_1}^1 \cdots a_{k_d,\ell_d}^d \, f_{(\ell_1,\dots,\ell_d)} \,,$$

where $A^1, \dots, A^d$ are the matrices describing the univariate scheme, based on the knot sequences ξ_ν and the degrees n_ν. This amounts to an application of the univariate scheme in each of the components with the indices corresponding to the other components held fixed. More precisely, the B-spline coefficients are computed in d steps:

$$f = f^0 \to f^1 \to \cdots \to f^d = c \,,$$

where

$$f^\nu(\dots, k_\nu, \dots) = \sum_{\ell_\nu} a_{k_\nu,\ell_\nu}^\nu \, f^{\nu-1}(\dots, \ell_\nu, \dots)$$

for $\nu = 1, \dots, d$.

The tensor product formalism includes schemes involving the solution of a linear system. In this case, A is the inverse of the system matrix Q. Accordingly, to compute f^ν, the linear system

$$\sum_{k_\nu} q_{\ell_\nu,k_\nu}^\nu f^\nu(k_1, \dots, k_\nu, \ell_{\nu+1}, \dots, \ell_d) = f^{\nu-1}(k_1, \dots, k_{\nu-1}, \ell_\nu, \dots, \ell_d)$$

is solved for each of the indices $k_1, \dots, k_{\nu-1}$ and $\ell_{\nu+1}, \dots, \ell_d$. The superscript ν indicates that the system matrices may differ for the d coordinate directions.

In a first example, we discuss the construction of quasi-interpolants which are the basic tools for deriving error estimates.

☐ **Example:**

We consider a standard projector for quadratic splines

$$f \mapsto Qf = \sum_{k=0}^{m-1} c_k \, b_{k,\xi}^2 \in S_\xi^n, \quad c_k = Q_k f = -f_{2k}/2 + 2f_{2k+1} - f_{2k+2}/2 \,,$$

7.5. Approximation Methods

with double knots $\xi_1 = \xi_2$ and $\xi_m = \xi_{m+1}$ at the endpoints of the parameter interval, $f_\ell = f(t_\ell)$, and

$$t_0 = \xi_1, \ t_1 = (\xi_1 + \xi_2)/2, \ t_2 = \xi_2, \ldots.$$

Equivalently, $c = Af$ with

$$a_{k,2k} = w_0 = -\frac{1}{2}, \quad a_{k,2k+1} = w_1 = 2, \quad a_{k,2k+2} = w_2 = -\frac{1}{2},$$

as nonzero matrix entries. The corresponding bivariate scheme

$$f \mapsto \sum_{k_1=0}^{m_1-1} \sum_{k_2=0}^{m_2-1} (Q_{(k_1,k_2)} f) \, b_k$$

with b_k the biquadratic B-spline is obtained by forming products of the weights w_0, w_1, w_2, resulting in the functionals

$$Q_{(k_1,k_2)} f = \sum_{\alpha_1=0}^{2} \sum_{\alpha_2=0}^{2} w_{\alpha_1} w_{\alpha_2} f(t^1_{2k_1+\alpha_1}, t^2_{2k_2+\alpha_2}).$$

The points t^ν_ℓ are computed from the knot sequences ξ_ν, $\nu = 1, 2$, by inserting midpoints, as described above. In this fashion, we obtain a bivariate quasi-interpolant with uniformly bounded functionals Q_k, which reproduces all polynomials of coordinate degree $\leq (2,2)$ on the parameter rectangle $[\xi_{1,2}, \xi_{1,m_1}] \times [\xi_{2,2}, \xi_{2,m_2}]$.

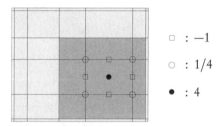

□ : -1

○ : $1/4$

● : 4

The figure shows the bivariate weights positioned at the evaluation points of f within the support of a B-spline. We note that in this particular example the weights are independent of the knot sequence. Moreover, since $Q_k f$ only depends on values of f in one grid cell, Q is a projector onto the bivariate spline space. □

In a second example, we illustrate the construction of multivariate spline interpolants.

□ **Example:**
The natural cubic spline interpolant $p = \sum_{k=0}^{m-1} c_k b_k$ with uniform knots $\xi_0 = -3h < -2h < \cdots < (M+2)h < (M+3)h = \xi_{m+3}$ for the data $f_0, \ldots, f_M$, $M = m-3$, minimizes the integral $\int_0^{Mh} |p''|^2$. Its coefficients c_k are determined by the linear system

$$\underbrace{\begin{pmatrix} -1 & 2 & -1 & & & \\ 1 & 4 & 1 & & & \\ & \ddots & \ddots & \ddots & & \\ & & 1 & 4 & 1 \\ & & -1 & 2 & -1 \end{pmatrix}}_{Q} \begin{pmatrix} c_0 \\ c_1 \\ \vdots \\ c_{m-2} \\ c_{m-1} \end{pmatrix} = \begin{pmatrix} 0 \\ 6f_0 \\ \vdots \\ 6f_M \\ 0 \end{pmatrix}.$$

The first and last equations describe the natural boundary conditions $p''(0) = 0 = p''(Mh)$. The remaining part of the system represents the interpolation conditions.

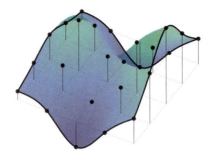

The figure shows a bicubic natural spline interpolant

$$p = \sum_{k_1=0}^{m-1} \sum_{k_2=0}^{m-1} c_k \, b_k$$

for data f_ℓ on a uniform square grid. It is computed as tensor product of the univariate scheme. First, we extend the $(M+1) \times (M+1)$-array f by zeros on each side to incorporate the natural boundary conditions, and we multiply the entries by 6. Denoting the resulting $m \times m$ array by F^0, the coefficients c_k satisfy

$$\sum_{k_2=0}^{m-1} \sum_{k_1=0}^{m-1} q_{\ell_2,k_2} q_{\ell_1,k_1} c_{k_1,k_2} = f^0_{\ell_1,\ell_2}.$$

Writing these equations in matrix form,

$$QCQ^t = F^0,$$

the two-stage solution procedure is apparent. We first solve $QF^1 = F^0$ for F^1 and then $CQ^t = F^1$ for C. □

With the aid of quasi-interpolants we can show that multivariate splines approximate with optimal order.

Error of Multivariate Spline Approximation

Assume that the local mesh ratio of the knot sequences ξ_ν of a spline space S^n_ξ is bounded by r. Then, any smooth function f can be approximated on the parameter hyperrectangle $R = D^n_\xi$ with optimal order:

$$|f(x) - p(x)| \leq c(n, r, d) \sum_{\nu=1}^{d} h_\nu(x)^{n_\nu+1} \|\partial_\nu^{n_\nu+1} f\|_{\infty,R} \quad \forall x \in R,$$

for some multivariate spline $p \in S^n_\xi$ and with $h_\nu(x)$ the width of the grid cell containing x in the νth coordinate direction.

7.5. Approximation Methods

We construct an appropriate approximation with a multivariate quasi-interpolant. According to the univariate theory, there exist standard projectors

$$f \mapsto Q^\nu f = \sum_{k=0}^{m_\nu-1} (Q_k^\nu f)\, b_k \in S_{\xi_\nu}^{n_\nu}$$

with

$$Q_k^\nu f = \sum_{\alpha=0}^{n_\nu} q_{k,\alpha}^\nu f(t_{\nu,k,\alpha}), \quad \sum_\alpha |q_{k,\alpha}^\nu| \le c_\nu(n_\nu),$$

and $t_{\nu,k,\alpha}$ in the support of $b_{k,\xi_\nu}^{n_\nu}$ intersected with the parameter hyperrectangle R. Using the tensor product formalism, we define a multivariate standard projector by

$$f \mapsto Qf = \sum_{k_1=0}^{m_1-1} \cdots \sum_{k_d=0}^{m_d-1} (Q_k f)\, b_k \in S_\xi^n$$

with

$$Q_k f = \sum_{\alpha_1=0}^{n_1} \cdots \sum_{\alpha_d=0}^{n_d} q_{k_1,\alpha_1}^1 \cdots q_{k_d,\alpha_d}^d f(t_k), \quad t_k = (t_{1,k_1,\alpha_1}, \ldots, t_{d,k_d,\alpha_d}) \in D_k = R \cap \operatorname{supp} b_k.$$

It is not difficult to see that Q reproduces polynomials of coordinate degree $\le n$ and that the functionals Q_k are uniformly bounded. In particular, the univariate estimate for the coefficients q_{k_ν,α_ν}^ν yields

$$|Q_k f| \le \left[\sum_{\alpha_1} \cdots \sum_{\alpha_d} |q_{k_1,\alpha_1}^1| \cdots |q_{k_d,\alpha_d}^d| \right] \|f\|_{\infty,D_k}, \quad [\ldots] \le \prod_\nu c_\nu(n_\nu) = \|Q\|.$$

After these preliminary considerations, the proof of the error estimate proceeds in a familiar fashion. For a given point $x \in R$, we denote by D_x the intersection of R with the hyperrectangle formed by the union of the supports of all relevant B-splines b_k, $k \sim x$, and by p_x the orthogonal projection of f onto $\mathbb{P}^n(D_x)$. Then, with $p = Qf$, we have

$$|f(x) - p(x)| \le |f(x) - p_x(x)| + |(Qp_x)(x) - (Qf)(x)|$$

since $Qp_x = p_x$. By the boundedness of the functionals Q_k, the second summand is

$$\le \sum_{k \sim x} |Q_k(p_x - f)|\, b_k(x) \le \|Q\| \|p_x - f\|_{\infty,D_x} \sum_k b_k(x)$$

$$\le c(n,d) \sum_{\nu=1}^d h_\nu(D_x)^{n_\nu+1} \|\partial_\nu^{n_\nu+1} f\|_{\infty,D_x}$$

by the error estimate for multivariate polynomial approximation and with $h_\nu(D_x)$ the width of D_x in the νth coordinate direction. Obviously, this estimate holds for the first summand $|f(x) - p_x(x)|$ as well. Finally, because of the boundedness of the local mesh ratio,

$$h_\nu(D_x) \le (1 + 2r + \cdots + 2r^{n_\nu}) h_\nu(x)$$

completing the proof.

As in the univariate case, discussed in Section 5.3, we have only considered approximation in the maximum norm. Using essentially the same well-known formalism, estimates in L_p-norms are possible. Moreover, bounds for derivatives can be derived as well.

7.6 ▪ Hierarchical Bases

Because of their local support, B-splines are ideally suited for adaptive methods. However, in several variables, refinement of knot sequences does not provide optimal local flexibility. Adding a knot has a global effect. Nevertheless, a natural simple strategy exists. With nested spline spaces on different grids the approximation power can be increased in the neighborhood of critical areas. For a precise description, it is helpful to consider a simple univariate example first.

☐ **Example:**

The figure shows a typical situation. The function

$$x \to \sin\left(x^{0.35} e^{0.22x}\right), \quad x \in D = [0, 10],$$

cannot be well approximated with uniform splines. On the left of the parameter interval D, the derivative is infinite, and the oscillatory behavior increases towards the right. In both areas local refinement is needed. We illustrate this by adaptively constructing a piecewise linear approximation with error less than a prescribed tolerance.

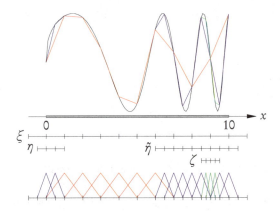

We start with an initial interpolating spline

$$p = \sum_k c_{\xi,k} b^1_{k,\xi}, \quad \xi : -1, \ldots, 11$$

(red polygon). Then, we use additional B-splines with finer grid width $h = 1/2$ on the intervals $[0, 1]$ and $[6, 10]$, where the error exceeds the given tolerance (0.3 in the concrete example). We add updates

$$q = \sum_k c_{\eta,k} b^1_{k,\eta}, \quad \eta : -0.5, 0, 0.5, 1,$$

$$\tilde{q} = \sum_k c_{\tilde{\eta},k} b^1_{k,\tilde{\eta}}, \quad \tilde{\eta} : 6, 6.5, \ldots, 10.5$$

(blue polygons). On the interval $[8.5, 9.5]$, one further refinement is needed. Adding a linear combination r of the B-splines $b^1_{k,\zeta}, k = 0, 1, 2$, with $\zeta : 8.5, 8.75, 9, 9.25, 9.5$ (green polygon), the sum

$$p + q + \tilde{q} + r$$

achieves the desired accuracy.

7.6. Hierarchical Bases

The adaptive refinement can be described by a tree structure, having the knot sequences as nodes. A basis for the resulting hierarchical spline space is obtained by discarding those B-splines which can be represented on $D = [0, 10]$ as linear combinations of B-splines on finer grids. For the concrete example, the hierarchical basis depicted at the bottom of the figure consists of

$$b^1_{1,\xi}, \ldots, b^1_{6,\xi}, \quad b^1_{0,\eta}, b^1_{1,\eta}, \quad b^1_{0,\tilde{\eta}}, \ldots, b^1_{4,\tilde{\eta}}, b^1_{6,\tilde{\eta}}, b^1_{7,\tilde{\eta}}, \quad b^1_{0,\zeta}, b^1_{1,\zeta}, b^1_{2,\zeta},$$

noting that labeling of the knots starts from 0. □

The generalization of the natural refinement principle to several variables is clear. However, some additional notation is useful. For knot sequences ξ, we denote by $[\xi]$ the (smallest) hyperrectangle enclosing the grid points:

$$[\xi] = [\xi_{1,0}, \xi_{1,m_1+n_1}] \times \cdots \times [\xi_{d,0}, \xi_{d,m_d+n_d}].$$

Moreover, we say that η is a refinement of ξ if $[\eta] \subseteq [\xi]$ and, for all $\nu = 1, \ldots, d$, the knots $\xi_{\nu,\ell} \in [\eta_\nu]$ form a subsequence of η_ν. The figure shows an example.

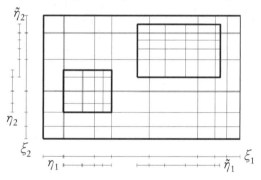

The following definition of hierarchical spline spaces is related to the approach by Höllig in the context of finite elements and the more recent description of hierarchical bases proposed by Lyche, Dokken, and Petterson.

Hierarchical Splines

A hierarchical spline space $S^n_\Xi(D)$ is spanned by B-splines $b^n_{k,\xi}$ from knot sequences ξ, which are nodes of a tree Ξ. It is assumed that each of the children η is a refinement of its parent ξ, and the enclosing hyperrectangles $[\eta]$ do not intersect. Moreover, if $D \cap \operatorname{supp} b^n_{k,\xi}$ is contained in $[\eta]$, then this B-spline must be representable as linear combination of B-splines from η on D.

A basis for $S^n_\Xi(D)$ consists of those relevant B-splines for D,

$$b^n_{k,\xi}, \quad k \in K_\xi, \xi \in \Xi,$$

which are nonzero at a point in the interior of D outside of the hyperrectangles $[\eta]$ for any of the children η of ξ.

Obviously, the B-splines in the basis span $S^n_\Xi(D)$. For, if $b^n_{k,\xi}$ does not belong to the basis ($k \notin K_\xi$), either it is not relevant ($b^n_{k,\xi} = 0$ on D), or its support within D is contained in $[\eta]$ for some child η of ξ. In the latter case, by assumption, $b_{k,\xi}^n$ is representable as linear combination of the B-splines $b_{k,\eta}^n$.

To check the linear independence of the basis elements, we assume that

$$p = \sum_{\xi \in \Xi} \sum_{k \in K_\xi} c_{\xi,k} b_{k,\xi}^n(x) = 0 \quad \forall x \in D.$$

Let $D(\xi^*)$ be the intersection of the interior of D with the complements of the hyperrectangles $[\eta]$ for all children of the root ξ^* of Ξ. Then, all B-splines $b_{k,\xi}^n$ with $\xi \neq \xi^*$ vanish on $D(\xi^*)$. Since the B-splines b_{k,ξ^*}^n in the basis ($k \in K_{\xi^*}$), by assumption, all have some nontrivial support on $D(\xi^*)$, they are linearly independent on this set. Consequently, their coefficients in the linear combination p are zero. We now repeat this argument for the B-splines associated with the children η of ξ^*, and conclude that their coefficients are zero, too. In this way, we can traverse the entire tree, completing the proof.

Let us comment on the assumptions on the tree Ξ. Typically, the knot sequences arise as the result of an adaptive procedure controlled by some error measure. Therefore, it is natural to require that children are refinements of their parent. Moreover, if $[\eta] \cap [\tilde{\eta}] \neq \emptyset$ for two children of ξ, it is reasonable to combine these knot sequences. Hence, the strict separation of knot sequences on the same level of Ξ, which has obvious programming advantages, is no severe restriction. The second assumption is automatically satisfied if $[\eta]$ is a subset of the interior of D and if all common knots in ξ_ν and η_ν have the same multiplicities. The condition essentially restricts the admissible choices of refinements at the boundary. We do not want to keep coarse grid B-splines in the basis if their support within D is contained in $[\eta]$. For example, in the figure at the beginning of this section, choosing $\eta: 0, 0.5, 1$ would require to keep $b_{0,\xi}^1$ in the basis since this coarse grid B-spline is no longer representable in terms of the B-splines $b_{k,\eta}^1$ on $D = [0, 10]$.

☐ **Example:**

The figure below gives an example of a hierarchical spline space $S_\Xi^n(D)$ for bilinear splines ($n = (1,1)$) with uniform knots on a heart-shaped domain $D \subset \tilde{R} = [2,6] \times [2,5]$. We have shown the grid corresponding to the knot sequences and marked the B-splines in the basis at their knot averages. Different markers are used to distinguish the levels of refinement. The tree Ξ is depicted on the right of the figure.

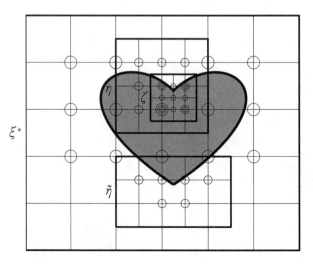
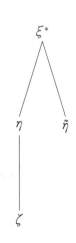

7.6. Hierarchical Bases

153

The thick lines are the boundaries of the enclosing rectangles for the knot sequences. In accordance with the definition of a spline space, $[\xi^*]$ extends $n_v = 1$ layers of grid cells beyond the bounding rectangle R for the domain D. $\qquad\square$

Hierarchical splines were defined in great generality. However, in practice, uniform knot sequences ξ_v and equal coordinate degrees n_v are used almost exclusively. This means the hierarchical basis for $S_\Xi^n(D)$ consists of scaled translates

$$x \mapsto b_{k,\xi}^n(x) = b^n((x - (\xi_{1,k_1}, \ldots, \xi_{d,k_d}))/h), \quad k_v \sim \xi_v,$$

of the standard uniform B-spline b^n. The grid width h is halved for each refinement step. Of course, this is an ideal situation from a programming point of view, permitting an extremely simple data structure. We describe this in more detail.

For each node ξ of the tree Ξ, we just store the grid width h_ξ, the hyperrectangle $[\xi]$, and the coefficient array

$$c_{\xi,k}, \quad k \sim \xi,$$

corresponding to the B-splines from ξ. In a familiar way, we do not store index lists to identify the coefficients $c_{\xi,k}$, $k \in K_\xi$, of the B-splines in the basis. Instead, we use a flag or 0 as value for irrelevant coefficients. The tree structure is recorded by pointers to parent and children in the usual way.

As an illustration, we list the data for a spline in the hierarchical space of the above example. For the nodes $\xi^*, \eta, \tilde{\eta}, \zeta$, the following information is stored:

$\xi^*: \to \eta, \tilde{\eta}$

$$h = 1, \quad [\xi^*] = [1,7] \times [1,6], \quad c_{\xi^*} = \begin{pmatrix} 0 & \times & \times & \times \\ 0 & \times & \times & \times \\ 0 & \times & \times & 0 \\ 0 & \times & \times & \times \\ 0 & \times & \times & \times \end{pmatrix},$$

$\eta: \leftarrow \xi^*$

$$h = 1/2, \quad [\eta] = [3,5] \times [3.5,5.5], \quad c_\eta = \begin{pmatrix} \times & \times & \times \\ \times & \times & \times \\ \times & \times & \times \end{pmatrix},$$

$\tilde{\eta}: \leftarrow \xi^*, \to \zeta$

$$h = 1/2, \quad [\tilde{\eta}] = [3,5.5] \times [1.5,3], \quad c_{\tilde{\eta}} = \begin{pmatrix} 0 & \times \\ \times & \times \\ \times & \times \\ 0 & \times \end{pmatrix},$$

$\zeta: \leftarrow \tilde{\eta}$

$$h = 1/4, \quad [\zeta] = [3.75, 4.75] \times [3.75, 4.75], \quad c_\zeta = \begin{pmatrix} \times & \times & \times \\ \times & \times & \times \\ \times & \times & \times \end{pmatrix}.$$

In the coefficient matrices, irrelevant entries are marked with 0. Counting the remaining entries, we see that $\dim S_\Xi^{(1,1)}(D) = 14 + 9 + 6 + 9 = 38$ in this example.

Note that the positions of the relevant entries in the coefficient matrices match the positions of the B-spline centers in the figure only after rotation. This is because the x_1-axis is horizontal while the first matrix index refers to the (vertical) row positions.

Chapter 8

Surfaces and Solids

Methods for modeling and approximation of surfaces and solids are similar to the techniques used for curves. Parametrizations are obtained by replacing the scalar coefficients of multivariate splines or polynomials in Bernstein form by control points in $\mathbb{R}^3$. Because of their flexibility and accuracy, such representations have become the method of choice for CAD/CAM applications.

In Sections 8.1 and 8.2, we describe Bézier and spline parametrizations. Subdivision strategies, which are essential for numerical processing and rendering of surfaces, are discussed in Section 8.3. We then consider the construction of surfaces from curve networks in Section 8.4. Finally, Section 8.5 is devoted to the representation of solids.

8.1 ▪ Bézier Surfaces

Classical surface representations use polynomials. As for curves, the Bernstein basis is particularly well suited for design applications. Moreover, this basis has favorable algorithmic and numerical properties.

Bézier Patch

A Bézier patch of coordinate degree $\leq n$ has a parametrization

$$(t_1, t_2) \mapsto (r_1(t), r_2(t), r_3(t)), \quad 0 \leq t_v \leq 1,$$

in terms of bivariate Bernstein polynomials:

$$r = \frac{\sum_{k_1=0}^{n_1} \sum_{k_2=0}^{n_2} (c_k w_k) b_k^n}{\sum_{k_1=0}^{n_1} \sum_{k_2=0}^{n_2} w_k b_k^n}, \quad b_k^n(t) = b_{k_1}^{n_1}(t_1) b_{k_2}^{n_2}(t_2),$$

with control points $c_k \in \mathbb{R}^3$ and weights $w_k > 0$.

155

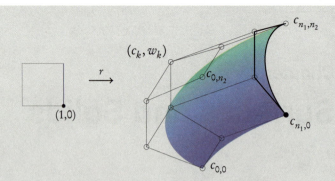

The control net C qualitatively describes the shape of the surface with additional design flexibility provided by the weights. In particular, the corners of the patch coincide with the control points

$$c_{0,0}, c_{n_1,0}, c_{0,n_2}, c_{n_1,n_2},$$

and the boundary consists of the four rational Bézier curves corresponding to the boundary polygons of the control net.

The assertions about the patch boundaries are a consequence of the endpoint interpolation property of Bernstein polynomials:

$$b_j^m(0) = \delta_{j,0}, \quad b_j^m(1) = \delta_{j,m}.$$

Considering, e.g., the corner with parameters $(t_1, t_2) = (0,0)$, $b_{0,0}^n(t) = 1$ and all other bivariate Bernstein polynomials vanish at t. Hence,

$$r(0,0) = \frac{c_{0,0} w_{0,0}}{w_{0,0}} = c_{0,0}.$$

Similarly, for $t_2 = 0$, the sum over k_2 reduces to the term with $k_2 = 0$, i.e.,

$$r(t_1, 0) = \frac{\sum_{k_1=0}^{n_1} (c_{k_1,0} w_{k_1,0}) b_{k_1}^{n_1}(t_1)}{\sum_{k_1=0}^{n_1} w_{k_1,0} b_{k_1}^{n_1}(t_1)}$$

since $b_0^{n_2}(0) = 1$ and $b_1^{n_2}(0) = \cdots = b_{n_2}^{n_2}(0) = 0$. This shows that the boundary corresponding to $t_2 = 0$ is a rational Bézier curve with control points $c_{0,0}, \ldots, c_{n_1,0}$ and weights $w_{0,0}, \ldots, w_{n_1,0}$.

The other cases are handled similarly.

□ **Example:**

If all weights w_k are equal to 1, we obtain a polynomial Bézier patch as a special case of a rational parametrization:

$$p = \sum_{k_1=0}^{n_1} \sum_{k_2=0}^{n_2} c_k b_k^n.$$

8.1. Bézier Surfaces

As an example, we consider a bicubic patch, which is determined by 16 control points c_k, $0 \le k_1, k_2 \le 3$.

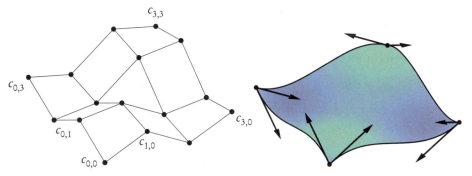

An alternative description is provided by derivative data at the four corner points. By the differentiation formula for Bézier curves (which form the patch boundaries),

$$\frac{\partial p(s,0)}{\partial s} = \frac{\partial}{\partial s} \sum_{j=0}^{3} c_{j,0} b_j^3(s) = 3 \sum_{j=0}^{2} (c_{j+1,0} - c_{j,0}) b_j^2(s),$$

which equals $3(c_{1,0} - c_{0,0})$ for $s = 0$. Similarly, $(\partial p(0,s)/\partial s)_{s=0} = 3(c_{0,1} - c_{0,0})$. Hence, the tangent plane at $p(0,0)$ is determined by the triangle of the three control points at the corner if they are not collinear. Clearly, by symmetry, the analogous statement holds for the other patch corners. This leaves the four middle control points, the so-called twist points, as degrees of freedom. They correspond to the mixed partial derivatives. For example,

$$\partial_1 \partial_2 p(t_1, t_2)|_{t_v = 0} = 9(c_{1,1} - c_{1,0} - c_{0,1} + c_{0,0}).$$

The figure below gives a few examples of the design possibilities with bicubic patches.

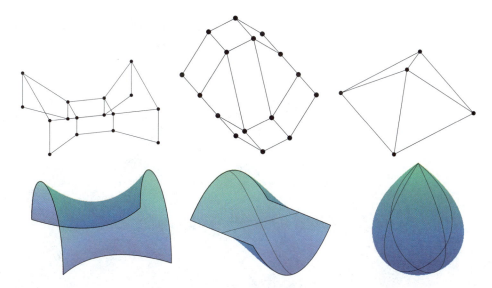

The construction on the right is slightly unusual; the tip of the "rain drop" has multiplicity 12 comprising all boundary control points of the bicubic patch. It is a singular point of the surface without tangent plane. □

A primary motivation for using rational parametrizations is that standard shapes can be represented exactly.

☐ **Example:**

A part of the unit sphere can be represented by a biquadratic Bézier patch. For the segment with the parametrization

$$x_1 = \cos(\varphi)\sin(\vartheta),\ x_2 = \sin(\varphi)\sin(\vartheta),\ x_3 = \cos(\vartheta),\quad \varphi \in [\varphi_0, \varphi_1],\ \vartheta \in [\vartheta_0, \vartheta_1],$$

in spherical coordinates, the homogeneous control points $(w_k c_{k,1}, w_k c_{k,2}, w_k c_{k,3} \,|\, w_k), 0 \leq k_1, k_2 \leq 2$, are

$$\begin{pmatrix} \cos(((2-k_1)\varphi_0 + k_1\varphi_1)/2)\sin(((2-k_2)\vartheta_0 + k_2\vartheta_1)/2) \\ \sin(((2-k_1)\varphi_0 + k_1\varphi_1)/2)\sin(((2-k_2)\vartheta_0 + k_2\vartheta_1)/2) \\ \dfrac{w_{\varphi,k_1}\cos(((2-k_2)\vartheta_0 + k_2\vartheta_1)/2)}{w_{\varphi,k_1} w_{\vartheta,k_2}} \end{pmatrix}^t$$

$$w_\varphi = (1, \cos((\varphi_1 - \varphi_0)/2), 1),\quad w_\vartheta = (1, \cos((\vartheta_1 - \vartheta_0)/2), 1).$$

The figure shows the control net and the sphere segment for $\varphi \in [0, \pi/6]$, $\vartheta \in [\pi/8, 3\pi/8]$.

☐

A single Bézier patch has only limited design capabilities. However, Bézier patches can be easily connected, yielding piecewise rational surfaces of arbitrary topological form.

Bézier Surface

A Bézier surface consists of Bézier patches which match along patch boundaries. This means that the corresponding boundary curves have parametrizations with the same control points and weights.

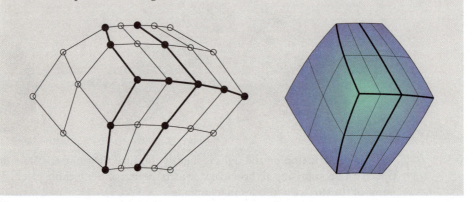

8.1. Bézier Surfaces

Bézier surfaces have no built-in smoothness. It is not straightforward to achieve high order differentiability across patch boundaries. Merely conditions for a continuous surface normal have a relatively simple form. For polynomial patches p, $\tilde{p}$ with equal coordinate degrees n, sufficient for tangent plane continuity is that the boundary control points are averages of the neighboring control points, as is illustrated in the figure.

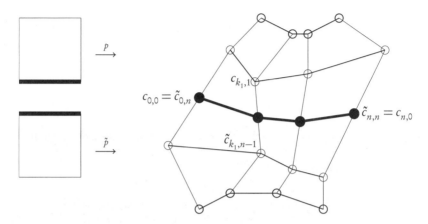

The sufficient smoothness condition

$$c_{k_1,0} = \tilde{c}_{k_1,n} = \frac{1}{2}\left(c_{k_1,1} + \tilde{c}_{k_1,n-1}\right)$$

implies that the normal derivatives match along the common boundary of the two polynomial patches. This is easily checked. By the endpoint interpolation property of Bézier curves we have

$$\partial_2 \left(\sum_{k_2} \left[\sum_{k_1} c_k b_{k_1}^n(t_1) \right] b_{k_2}^n(t_2) \right)_{|t_2=0} = n \sum_{k_1} \left(c_{k_1,1} - c_{k_1,0} \right) b_{k_1}^n(t_1).$$

Similarly, the normal derivative of the patch $\tilde{p}$ along the common boundary is

$$\partial_2 \tilde{p}(t_1, t_2)_{|t_2=1} = n \sum_{k_1} \left(\tilde{c}_{k_1,n} - \tilde{c}_{k_1,n-1} \right) b_{k_1}^n(t_1).$$

The smoothness condition implies that the normal derivatives have the same control points and are thus identical. Since the tangential derivatives also match, the neighboring patches have the same tangent planes at all common boundary points.

For a regular patch structure, i.e., if $m = 4$ patches meet at every interior patch corner, continuously differentiable Bézier surfaces can be easily constructed. For each patch, the interior control points may be chosen arbitrarily. Then, the control points of the boundary curves are uniquely determined by the smoothness conditions. This does not lead to inconsistencies at the patch corners. With the notation of the figure below,

$$2e_\nu = f_{\nu-1} + f_\nu \ \forall \nu \quad \Longrightarrow \quad 2c = e_\nu + e_{\nu+2} \ \forall \nu$$

with the indices taken modulo $m = 4$.

Unfortunately, for irregular points ($m \neq 4$), the sufficient smoothness conditions around a corner admit only the trivial solution

$$c = e_\nu = f_\nu.$$

The simple remedy proposed by Reif is to allow such multiple control points, as shown in the right figure for $m = 3$ and degree $n = (3,3)$.

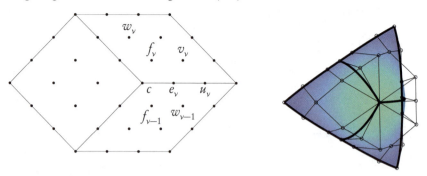

For the resulting singularly parametrized patches, the tangent plane is now determined by the next layer of control points. Requiring that v_ν, w_ν ($\nu = 0, \ldots, m-1$) are coplanar, the condition

$$2u_\nu = v_\nu + w_{\nu-1} \pmod{m}$$

leads to continuously differentiable patch connections if no geometric singularities arise.

8.2 ▪ Spline Surfaces

Spline surfaces have become the standard in CAD/CAM applications. They include regular Bézier surfaces as a special case and provide very flexible and efficient representations. Moreover, because of the built-in smoothness, generally fewer control points are needed for accurate approximations.

> **Spline Surface**
> A spline surface has a parametrization
>
> $$D_\tau^n \ni (t_1, t_2) \mapsto (p_1(t), p_2(t), p_3(t))$$
>
> with components p_ν in a bivariate spline space S_τ^n of degree (n_1, n_2) and with knot sequences $\tau_\nu : \tau_{\nu,0}, \ldots, \tau_{\nu, m_\nu + n_\nu}$ ($\nu = 1, 2$). This means
>
> $$p(t) = \sum_{k_1=0}^{m_1-1} \sum_{k_2=0}^{m_2-1} c_k\, b_{k,\tau}^n(t), \quad \tau_{\nu,n_\nu} \leq t_\nu \leq \tau_{\nu, m_\nu},$$
>
> with control points $(c_{k,1}, c_{k,2}, c_{k,3}) \in \mathbb{R}^3$.
>
>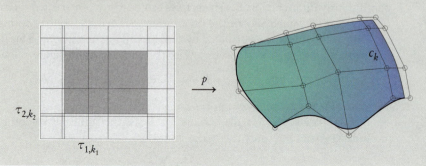

8.2. Spline Surfaces

> Connecting neighboring control points, we obtain a control net for the surface, which provides a qualitative model of the geometric shape.
> Closed (partially closed) surfaces are modeled by periodic parametrizations in both (one) parameter directions.

More generally, we can define rational spline surfaces with parametrizations of the form

$$r = \frac{p}{q} = \frac{\sum_{k_1}\sum_{k_2}(c_k w_k)b_k}{\sum_{k_1}\sum_{k_2} w_k b_k}.$$

Such representations can be used, e.g., to combine Bézier patches providing exact descriptions of standard surfaces. One should note, however, that the tensor product structure as well as parametric smoothness limit the topological possibilities.

With the following examples, we illustrate different surface types.

□ **Example:**
We model a helical staircase with a spline surface of degree $n = (1,3)$ with uniform knots

$$\tau_1 : 0, 1, 2, 3, \quad \tau_2 : 0, \ldots, 14.$$

As is illustrated in the figure, to describe two turns of the surface we use the control points

$$c_k = ((1+k_1)\cos(\pi k_2/2), (1+k_1)\sin(\pi k_2/2), k_2), \quad k_1 = 0, 1, k_2 = 0, \ldots, 10,$$

corresponding to the parameter rectangle $D_\tau^{(1,3)} = [1,2] \times [3,11]$.

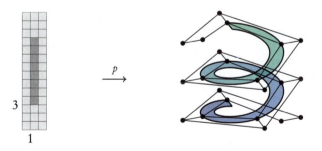

Since

$$b_{k,\tau}^n(t) = b^1(t_1 - k_1) b^3(t_2 - k_2),$$

with b^{n_ν} the univariate standard uniform B-spline, the parametrization has the form

$$p(t) = (2-t_1)\sum_{k_2} c_{0,k_2} b^3(t_2 - k_2) + (t_1 - 1)\sum_{k_2} c_{1,k_2} b^3(t_2 - k_2), \quad 1 \leq t_1 \leq 2.$$

Hence, p describes a ruled surface consisting of line segments, which connect the points on the two curves parametrized by $p(1, t_2)$ and $p(2, t_2)$ with $3 \leq t_2 \leq 11$. □

☐ **Example:**
The figure shows an approximation of a torus by a biquadratic uniform spline surface. The doubly 4-periodic parametrization has the form

$$t \mapsto p(t) = \sum_{k_1}\sum_{k_2} c_k\, b^n(t-k)$$

with b^n, $n = (2,2)$, the standard uniform biquadratic B-spline with support $[0,3]^2$, and with control points

$$\begin{array}{rcccccl}
c_{0,0} &=& (1,0,-1) & (0,1,-1) & (-1,0,-1) & (0,-1,-1) &= c_{0,3} \\
& & (3,0,-1) & (0,3,-1) & (-3,0,-1) & (0,-3,-1) & \\
& & (3,0,\ 1) & (0,3,\ 1) & (-3,0,\ 1) & (0,-3,\ 1), & \\
c_{3,0} &=& (1,0,\ 1) & (0,1,\ 1) & (-1,0,\ 1) & (0,-1,\ 1) &= c_{3,3}.
\end{array}$$

By periodicity,

$$c_{k_1,k_2} = c_{k_1+4,k_2} = c_{k_1,k_2+4}, \quad k_\nu \in \mathbb{Z}.$$

Accordingly, the control net consists of 4×4 quadrilaterals.

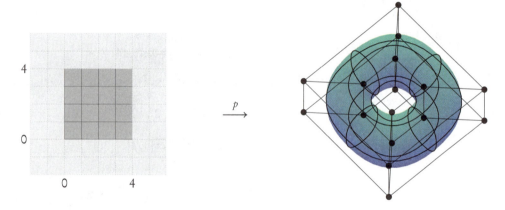

The periodicity is indicated by using dashed grid lines corresponding to knots $\tau_{\nu,k}$ outside the periodicity interval $[0,4)$.

Since $b^n(t-k) = b^2(t_1-k_1)b^2(t_2-k_2)$ and $b^2(1) = b^2(2) = 1/2$, we can easily compute points on the surface corresponding to values at the knots. For example,

$$p(0,0) = \frac{1}{4}(c_{-2,-2} + c_{-2,-1} + c_{-1,-2} + c_{-1,-1})$$
$$= \frac{1}{4}((-3,0,1) + (0,-3,1) + (-1,0,1) + (0,-1,1)) = (-1,-1,1),$$

noting that $c_{2,2} = c_{-2,-2}, c_{2,3} = c_{-2,-1}$, etc., by the periodicity conditions.

Similarly, one computes

$$p(2,0) = (-1,-1,-1), \quad p(0,2) = (1,1,1), \quad p(2,2) = (1,1,-1).$$

It follows that the approximate radii of the torus, which is centered at $(0,0,0)$, are $R = |p(2,2) - p(2,0)|/2 = \sqrt{2}$, $r = |p(2,2) - p(0,2)|/2 = 1$.

We recall that an exact representation with polynomial patches is not possible. ☐

8.2. Spline Surfaces

□ **Example:**

The figure below shows a partially closed surface. The parameters for the spline parametrization p, which is 3-periodic in the second variable, are

$$n = (2,3), \quad \tau_1 : 0, 1, 1, 2, 3, 3, 4, \quad \tau_2 : -3, -2, \ldots, 6$$

with the parameter rectangle $[1,3] \times [0,3)$. Accordingly, the control net consists of $4 \times 3 = 12$ control points given by

$$c_k = (k_1/3, \sin(k_1\pi/6)\sin(2k_2\pi/3), \sin(k_1\pi/6)\cos(2k_2\pi/3)), \quad k_1 = 0,1,2,3, \; k_2 = 0,1,2,$$

with $c_{k_1,k_2+3} = c_{k_1,k_2}$ for $k_2 \in \mathbb{Z}$.

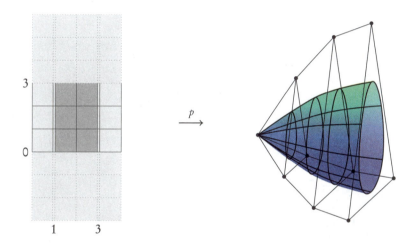

The multiple control point $c_{0,0} = c_{0,1} = c_{0,2}$ at the tip $p = (0,0,0)$ causes a singularity. There does not exist a tangent plane at this point. □

A spline surface can be converted to Bézier form by raising the multiplicity of each knot in the knot sequences τ_ν, $\nu = 1, 2$, to n_1 and n_2, respectively. The refined control net provides a closer approximation to the surface, in particular because of the endpoint interpolation property. Moreover, we can apply polynomial algorithms simultaneously to each surface segment.

The conversion is done via the tensor product formalism. We apply the curve algorithm first to the columns of the control point array, i.e., the control polygons formed by

$$c_{0,k_2}, \ldots, c_{m_1-1,k_2},$$

simultaneously for $k_2 = 0, \ldots, m_2 - 1$. The knot insertion is based on the knot sequence τ_1. Then, we process the rows of the modified array based on the knot sequence τ_2.

For smooth spline surfaces, rational parametrizations are not used very frequently. The algorithmic simplicity of piecewise polynomial parametrizations outweighs the additional flexibility provided by the weights. Moreover, standard objects, such as spheres or cylinders, cannot be represented with smooth regular spline parametrizations. Hence, a key feature of rational Bézier patches does not persist.

8.3 ▪ Subdivision Surfaces

Refinement of the control net via knot insertion provides a basic rendering scheme. It is particularly efficient for uniform knots. In this case, there are very simple and elegant formulas for simultaneous knot insertion at midpoints. To describe this in more detail, we first recall the corresponding algorithm for curves (see Section 6.3). The refined control polygon $\tilde{C}$ for a uniform parametrization

$$t \mapsto p(t) = \sum_k c_k b_{k,\tau}^n(t), \quad b_{k,\tau}^n(t) = b^n((t - \tau_k)/h)$$

with control polygon C and knot sequence τ is given by

$$\tilde{c}_k = \sum_i s_{k-2i} c_i, \quad s_j = 2^{-n} \binom{n+1}{j}.$$

The simple procedure is easily generalized to control nets for uniform spline surfaces via the tensor product formalism. We obtain

$$\tilde{c}_{k_1,k_2} = \sum_{i_1} \sum_{i_2} s_{k_1-2i_1} s_{k_2-2i_2} c_{i_1,i_2}$$

for a subdivision step $C \to \tilde{C}$ of surface meshes. Of particular importance is the cubic case with the subdivision mask

$$s = (1, 4, 6, 4, 1)/8.$$

In this concrete case, the expressions for the refined control points $\tilde{c}_k$ simplify slightly. Depending on the parity of (k_1, k_2) there are essentially three cases, as illustrated in the figure.

- Face points:

$$f = \frac{1}{4} \sum_{i \sim f} v_i,$$

  where the sum is taken over the four vertices of a quadrilateral (face) of the control polygon.

- Edge points:

$$e = (f_- + f_+ + v_- + v_+)/4,$$

  where $f_\pm$ are the two adjacent face points and $v_\pm$ are the vertices of the common edge.

- Vertex points:

$$\tilde{v} = \frac{9}{16} v + \frac{3}{32} \sum_{e \sim v} v_e + \frac{1}{64} \sum_{f \sim v} v_f,$$

  where v_e are the endpoints of the edges and v_f the remaining vertices of the faces meeting at v.

8.3. Subdivision Surfaces

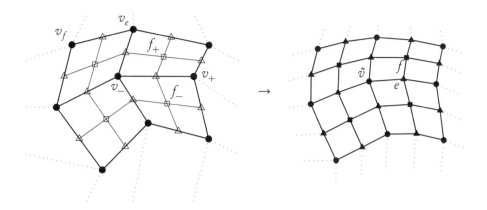

So far, the subdivision procedure applies to regular control nets only; i.e., every interior control point belongs to four quadrilaterals of the mesh. With a slight modification, the algorithm can be adapted to general quadrilateral control nets, which do not necessarily correspond to a spline parametrization.

Catmull–Clark Algorithm

A quadrilateral mesh can be subdivided by splitting each quadrilateral into four parts, according to the following rules.

(i) A new face point is the average of the four vertices of the enclosing quadrilateral.

(ii) A new edge point is the average of the two vertices of the edge and the two neighboring face points.

(iii) A vertex v of the coarse mesh is replaced by a convex combination of the vertices of the surrounding quadrilaterals with weights indicated in the figure:

$$v \leftarrow (1-\alpha-\beta)v + \frac{\alpha}{m}\sum_{\nu=1}^{m} v_{e,\nu} + \frac{\beta}{m}\sum_{\nu=1}^{m} v_{f,\nu},$$

where m denotes the number of edges incident at the vertex v.

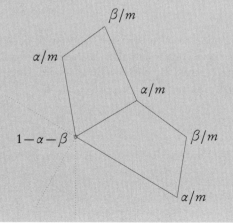

m	α	β
3	1/2	1/12
4	3/8	1/16
5	3/10	1/20
6	1/4	1/24

> The algorithm can be applied to quadrilateral meshes, which model closed surfaces or surfaces with boundary. In the latter case, the coarse boundary, i.e., one layer of quadrilaterals, is discarded after a subdivision step.
>
> For a regular quadrilateral mesh ($m = 4$), the algorithm is identical with uniform subdivision for bicubic spline surfaces.

The following example demonstrates the rapid convergence of the subdivision process. Usually a few steps already yield a visually smooth surface with appropriate rendering algorithms.

□ **Example:**
The figure shows several Catmull–Clark steps for the wire frame of a unit cube as the initial control net. In the limit, a sphere-like closed surface is obtained, approximated by an almost regular net. This means that the subdivisions lead to a regular tensor product control net except for isolated points, corresponding to the vertices of the initial cube. These six points have three instead of four neighbors.

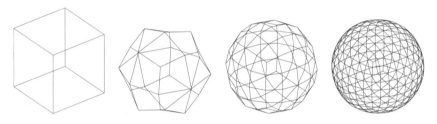

□

The sequence of meshes appears to converge to a smooth surface. However, analyzing the differentiability properties of the limit is rather delicate. If no geometric singularities occur, then the limit surface is twice continuously differentiable except for the limits of isolated irregular vertices. At these points, merely the surface normal is continuous, the curvatures can become infinite (see the book by Peters and Reif for a comprehensive analysis).

8.4 ▪ Blending

The classical scheme for constructing surfaces from curve networks is based on linear interpolation. For each curved quadrilateral of the mesh, the bounding curves are blended together to form a surface patch. By construction, neighboring patches match, and this technique yields a globally continuous interpolating surface.

> **Coons Patches**
> Coons's method approximates a surface patch parametrized by
>
> $$t \mapsto p(t_1, t_2), \quad 0 \le t_v \le 1,$$
>
> by linearly interpolating its four boundary curves ($t_v = 0$ or $t_v = 1$).

8.4. Blending

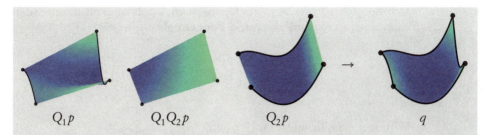

$Q_1 p$ $\qquad$ $Q_1 Q_2 p$ $\qquad$ $Q_2 p$ $\qquad$ q

Denoting by Q_ν the interpolation operator in the νth direction, the interpolating Coons patch has the parametrization

$$q = Q_1 p + Q_2 p - Q_1 Q_2 p.$$

More explicitly,

$$\begin{aligned} q(t_1, t_2) = & [(1-t_1)p(0,t_2) + t_1 p(1,t_2)] + [(1-t_2)p(t_1,0) + t_2 p(t_1,1)] \\ & - [(1-t_1)(1-t_2)p(0,0) + (1-t_1)t_2 p(0,1) + t_1(1-t_2)p(1,0) + t_1 t_2 p(1,1)]. \end{aligned}$$

We have to show that

$$p(t_1, t_2) = q(t_1, t_2)$$

if either t_1 or t_2 is 0 or 1. This is easily verified directly from the explicit formula for the Coons patch. Alternatively, we can use that the operators Q_1 and Q_2 are projectors and commute:

$$Q_\nu Q_\nu = Q_\nu, \quad Q_1 Q_2 = Q_2 Q_1.$$

This implies that

$$Q_1 q = Q_1(Q_1 + Q_2 - Q_1 Q_2) p = Q_1 p,$$

which shows that q and p have the same boundary curves in the t_2-direction. Similarly, $Q_2 q = Q_2 p$ implies that $p(t_1, 0) = q(t_1, 0)$ and $p(t_1, 1) = q(t_1, 1)$.

☐ **Example:**

We can use Coons's method to construct a polynomial Bézier patch parametrized by

$$p = \sum_{k_1=0}^{n_1} \sum_{k_2=0}^{n_2} c_k b_k^n, \quad 0 \leq t_\nu \leq 1,$$

from the Bézier curves making up its boundary.

The boundary curves are parametrized by

$$p(t_1, \nu) = \sum_{k_1=0}^{n_1} c_{k_1, \nu n_2} b_{k_1}^{n_1}(t_1), \quad p(\nu, t_2) = \sum_{k_2=0}^{n_2} c_{\nu n_1, k_2} b_{k_2}^{n_2}(t_2),$$

with $\nu = 0, 1$. Combining these Bézier representations with the identities

$$s = \sum_{j=0}^{m} \frac{j}{m} b_j^m(s), \quad 1 - s = \sum_{j=0}^{m} \frac{m-j}{m} b_j^m(s),$$

according to the formulas for the Coons patch, we obtain explicit expressions for the inner control points of the Bézier patch. For example,

$$Q_1 p = \sum_{k_1=0}^{n_1} \sum_{k_2=0}^{n_2} \left[\frac{n_1-k_1}{n_1} c_{0,k_2} + \frac{k_1}{n_1} c_{n_1,k_2} \right] \underbrace{b_{k_1}^{n_1}(t_1) b_{k_2}^{n_2}(t_2)}_{b_k^n(t)}.$$

With analogous formulas for $Q_2 p$ and $Q_1 Q_2 p$, we have

$$c_k = \left[\frac{n_1-k_1}{n_1} c_{0,k_2} + \frac{k_1}{n_1} c_{n_1,k_2} \right] + \left[\frac{n_2-k_2}{n_2} c_{k_1,0} + \frac{k_2}{n_2} c_{k_1,n_2} \right]$$
$$- \left[\frac{n_1-k_1}{n_1} \frac{n_2-k_2}{n_2} c_{0,0} + \frac{n_1-k_1}{n_1} \frac{k_2}{n_2} c_{0,n_2} + \frac{k_1}{n_1} \frac{n_2-k_2}{n_2} c_{n_1,0} + \frac{k_1}{n_1} \frac{k_2}{n_2} c_{n_1,n_2} \right]$$

for $0 < k_1 < n_1, 0 < k_2 < n_2$.

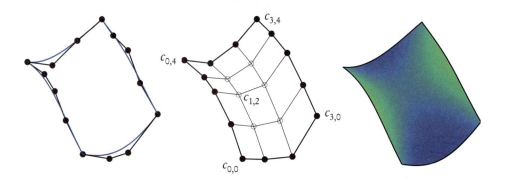

The figure illustrates the procedure for degree $n = (3,4)$. From the control polygons of the boundary curves we determine the six inner control points shown in the middle figure. For example,

$$c_{1,2} = \left[\frac{2}{3} c_{0,2} + \frac{1}{3} c_{3,2} \right] + \left[\frac{2}{4} c_{1,0} + \frac{2}{4} c_{1,4} \right]$$
$$- \left[\frac{2}{3} \frac{2}{4} c_{0,0} + \frac{2}{3} \frac{2}{4} c_{0,4} + \frac{1}{3} \frac{2}{4} c_{3,0} + \frac{1}{3} \frac{2}{4} c_{3,4} \right].$$

The resulting blended patch is shown on the right. □

If the boundary of a patch consists of three or more than four boundary curves, Coons's method does not apply directly. Some pre-processing is required. For example, we can connect the midpoint of each curve with a point approximately in the center of the patch. The resulting curved quadrilaterals can then be interpolated with Coons's scheme. A drawback is the ambiguity in defining the auxiliary connecting curves, particularly since their proper choice is crucial for the shape of the blended surface. Finding good solutions with smooth patch connections is by no means straightforward (see the book by Hoschek and Lasser for a survey of classical techniques).

8.5 ▪ Solids

Solids can be defined in terms of their boundary, e.g., by a collection of spline or Bézier surfaces, which match continuously. While this so-called boundary representation is adequate for many applications, sometimes it is desirable to parametrize the entire interior of the solid. For example, isogeometric finite element techniques require such representations.

> **Spline Solids**
> A spline solid $S \subset \mathbb{R}^3$ is parametrized by trivariate splines:
>
> $$(t_1, t_2, t_3) \mapsto p(t) = \sum_{k_1=0}^{m_1-1} \sum_{k_2=0}^{m_2-1} \sum_{k_3=0}^{m_3-1} c_k b_{k,\tau}^n(t), \quad t \in D_\tau^n,$$
>
> with control points $(c_{k,1}, c_{k,2}, c_{k,3})$ and $D_\tau^n = [\tau_{1,n_1}, \tau_{1,m_1}] \times [\tau_{2,n_2}, \tau_{2,m_2}] \times [\tau_{3,n_3}, \tau_{3,m_3}]$.
>
>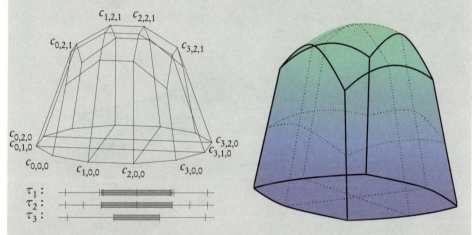
>
> The boundary of S consists of 6 spline surfaces, corresponding to the parameter values $t_\nu = \tau_{\nu,n_\nu}$, $t_\nu = \tau_{\nu,m_\nu}$ ($\nu = 1, 2, 3$).

As is depicted in the figure, for a regular parametrization with nonsingular Jacobian p', a spline solid normally retains the topological structure of the parameter cuboid. This means, the boundary of S has 12 edges and eight corners. However, as for surfaces, we can use partially periodic parametrizations. Moreover, we can combine different parametrizations. This considerably enhances the geometric possibilities, as is illustrated in the following examples.

☐ **Example:**
To model the tube-like shape shown in the figure, we use a periodic parametrization of degree $n_1 = 3$ in the first parameter variable t_1 with the knots

$$\tau_{1,0} = 0, 1, \ldots, 10 = \tau_{1,10}$$

relevant for the periodicity interval $[3, 7]$.

For the parameter variables t_2, t_3, we also choose uniform knots and degrees $n_2 = 2$, $n_3 = 1$, respectively:

$$\tau_2 : 0, 1, \ldots, 5, \qquad \tau_3 : 0, 1, 2, 3.$$

The control points are

$$c_k = (k_2, [\cosh(k_2-1)+k_3]\cos(k_1\pi/2), [\cosh(k_2-1)+k_3]\sin(k_1\pi/2))$$

with $k_2 = 0, 1, 2$ and $k_3 = 0, 1$.

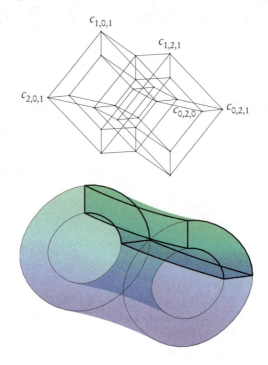

□

In our second example we show how to incorporate small features in a spline model.

□ **Example:**

The figure shows a gear wheel. For an exact representation of the outer circular boundary, we use a rational quadratic parametrization with

$$(c_k|w_k) : (1, 0, 0|1), (1, 1, 0|\sqrt{2}/2), (0, 1, 0|1)$$

as control points and weights for a quarter circle. Extending the arc with a simple linear parametrization in the radial and vertical direction, we obtain a rational Bézier parametrization of degree $n = (2, 1, 1)$ for a portion of the main part of the gear wheel.

We use a separate parametrization with the same degree for each of the teeth. Because of the triangular shape of the cross section, the three control points at the tip coalesce. The control points of the small arc are obtained by subdividing the parametrization of the quarter circle.

8.5. Solids

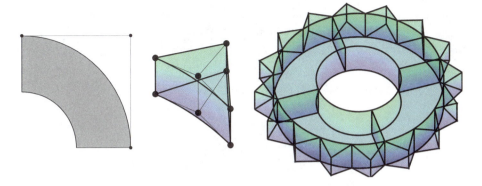

Since the circular boundary was represented exactly, matching the parametrizations did not present any problem. However, if a boundary surface is merely approximated, it can be difficult to combine solids without gaps or overlaps. □

A parametrization p can be used to compute physical characteristics of a solid S such as volume, center of gravity, and moments. For example,

$$\text{vol } S = \iiint_R \det p'(t)\, dt_1 dt_2 dt_3$$

if the determinant is positive. The columns of the Jacobi matrix p' are the partial derivatives of $(p_1, p_2, p_3)^t$ with respect to the parameters t_1, t_2, t_3. Hence, their components have degree at most (n_1-1, n_2, n_3), (n_1, n_2-1, n_3), (n_1, n_2, n_3-1), respectively, where n is the degree of p. It follows that $\det p'$ is a polynomial of degree $\leq 3n - (1,1,1)$ on each grid cell. Thus, we can compute the exact value of the integral via Gauß cubature with $\lceil 3n_\nu/2 \rceil$ points in the νth coordinate direction.

Another application is the construction of isogeometric finite element bases, as will be discussed in Section 9.3.

Chapter 9

Finite Elements

Finite element techniques are the method of choice for solving partial differential equations on domains with curved boundaries. A key advantage is the geometric flexibility of the basis functions. Classical methods are based on low degree piecewise polynomial approximations on triangulations of the simulation domain. Already for moderate tolerances, the discrete systems become fairly large, especially in three dimensions. Using B-splines instead yields highly accurate numerical solutions with relatively few parameters, which can be computed very efficiently.

The basic finite element idea is briefly described in Section 9.1. In Sections 9.2 and 9.3, we introduce two different types of bases: weighted B-splines on uniform grids and isogeometric elements parametrized over hyperrectangles. The assembly of the discrete equations is completely analogous for both approaches, as is outlined in Section 9.4. Finally, in Section 9.5, the performance of B-spline-based finite elements is illustrated for two typical applications.

9.1 ▪ Ritz–Galerkin Approximation

Many physical or engineering problems admit a variational formulation. This means that the function u describing a phenomenon or process on a bounded domain $D \subset \mathbb{R}^d$ minimizes an energy functional over a Hilbert space H, which incorporates boundary conditions if necessary:

$$Q(u) \to \min, \quad u \in H.$$

A finite element approximation

$$u_h = \sum_{k \in K} u_k B_k$$

minimizes Q over a finite dimensional subspace $\mathbb{B}_h = \operatorname{span}_k B_k$ of H:

$$Q(u_h) = \min_{v_h \in \mathbb{B}_h} Q(v_h),$$

where the discretization parameter h usually denotes a grid width. The set K contains the admissible indices k which need not be integers. Clearly, the choice of $\mathbb{B}_h$ as well as of the basis functions or finite elements B_k is crucial for the accuracy and the efficiency of the resulting method. An enormous number of different possibilities is available. As we will see in the following sections, B-splines are an excellent choice!

173

An important special case covering numerous applications is that of quadratic energy functionals:

$$Q(u) = \frac{1}{2}a(u,u) - \lambda(u),$$

where $a(\cdot,\cdot)$ is a symmetric elliptic bilinear form and λ a continuous linear functional. Ellipticity guarantees existence of a unique minimizer u and requires that there exist positive constants $c_\pm$ such that

$$c_-\|v\|_H^2 \le a(v,v), \quad |a(v,\tilde{v})| \le c_+\|v\|_H\|\tilde{v}\|_H$$

for all $v, \tilde{v} \in H$.

For quadratic energy functionals, the computation of a finite element approximation is particularly simple.

Ritz–Galerkin Approximation

The finite element approximation

$$u_h = \sum_{k \in K} u_k B_k \in \mathbb{B}_h \subset H$$

for a quadratic energy functional with an elliptic bilinear form $a(\cdot,\cdot)$ and a bounded linear functional λ can be computed by solving the Ritz–Galerkin equations

$$GU = F, \quad U = (u_k)_{k \in K},$$

where

$$g_{k,k'} = a(B_k, B_{k'}), \quad f_k = \lambda(B_k).$$

The characterization of the Ritz–Galerkin approximation u_h is easily derived from the definitions. We have

$$Q(u_h) = \frac{1}{2}a\left(\sum_k u_k B_k, \sum_{k'} u_{k'} B_{k'}\right) - \lambda\left(\sum_k u_k B_k\right) = \frac{1}{2}U^{\mathrm{t}}GU - F^{\mathrm{t}}U$$

in view of the linearity of $a(\cdot,\cdot)$ and λ. Setting the gradient with respect to U to zero, we find that a minimizing vector U is a solution of $GU = F$. Uniqueness follows because G is positive definite. To this end, we note that ellipticity implies

$$U^{\mathrm{t}}GU = a(u_h, u_h) \ge c_-\|u_h\|_H^2 > 0$$

for $U \neq 0$.

☐ **Example:**

A basic example, which often serves as a test problem, is Poisson's functional with

$$Q(u) = \frac{1}{2}\int_D |\operatorname{grad} u|^2 - \int_D f u \, ;$$

9.1. Ritz–Galerkin Approximation

i.e.,
$$a(u,v) = \int_D (\operatorname{grad} u)^t \operatorname{grad} v, \quad \lambda(u) = \int_D fu,$$

are the bilinear form and the linear functional for $\mathcal{Q}$.

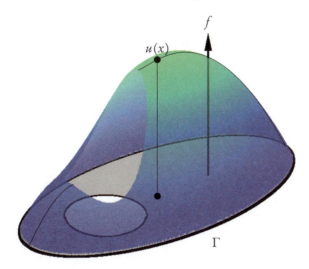

The figure illustrates a classical application. Minimizing $\mathcal{Q}$ yields the displacement $u(x)$ of an elastic membrane fixed at a subset Γ of the boundary ∂D under a vertical force with density f. Accordingly, the appropriate choice of the Hilbert space, which takes the boundary condition and the bilinear form into account, is the Sobolev space

$$H^1_\Gamma = \left\{ u : u_{|\Gamma} = 0, \int_D |u|^2, \int_D |\partial_\nu u|^2 < \infty \right\},$$

equipped with the norm

$$\|u\|_{1,D} = \left(\int_D u^2 + |\operatorname{grad} u|^2 \right)^{1/2}.$$

Using the calculus of variations, we find that a smooth minimizer u of $\mathcal{Q}$ satisfies Poisson's equation
$$-\Delta u = f \text{ in } D$$
($\Delta = \sum_\nu \partial_\nu^2$) with the boundary conditions
$$u = 0 \text{ on } \Gamma, \quad \partial_\perp u = 0 \text{ on } \partial D \setminus \Gamma,$$

where $\partial_\perp$ denotes the normal derivative. The first boundary condition is essential. It has to be incorporated into the finite element subspace $\mathbb{B}_h$. By contrast, the second, so-called natural, boundary condition is automatically satisfied by a minimizer of $\mathcal{Q}$. □

Despite the generality of the method, a simple error estimate for finite element approximations is possible. It compares the error $\|u - u_h\|_H$ with the error of the best approximation. Hence, the accuracy of finite elements can be judged independently of the differential equation.

Cea Lemma

The error of the Ritz–Galerkin approximation $u_h \in \mathbb{B}_h$ to the solution u of an elliptic problem with quadratic energy functional satisfies

$$\|u - u_h\|_H \leq (c_+/c_-) \inf_{v_h \in \mathbb{B}_h} \|u - v_h\|_H,$$

where $c_\pm$ are the ellipticity constants of the bilinear form.

The proof is based on the standard characterization of minimizers of quadratic functionals. For any $t \in \mathbb{R}$ and $v \in H$, $Q(u + tv) \geq Q(u)$, i.e.,

$$\frac{1}{2}a(u + tv, u + tv) - \lambda(u + tv) \geq \frac{1}{2}a(u, u) - \lambda(u).$$

Simplifying, using linearity and symmetry, we obtain the equivalent inequality

$$t\,[a(u, v) - \lambda(v)] + t^2\,a(v, v) \geq 0.$$

Since t is arbitrary, the expression in brackets must be zero:

$$a(u, v) = \lambda(v) \quad \forall v \in H.$$

Similarly, the minimality of u_h on $\mathbb{B}_h$ yields

$$a(u_h, v_h) = \lambda(v_h) \quad \forall v_h \in \mathbb{B}_h.$$

Setting $v = v_h = w_h$ and subtracting the above identities, we conclude that

$$a(u - u_h, w_h) = 0 \quad \forall w_h \in \mathbb{B}_h.$$

This orthogonality relation is the key in the derivation of Cea's inequality. It implies, with $w_h = u_h - v_h$,

$$a(u - u_h, u - u_h) = a(u - u_h, u - v_h).$$

By ellipticity, the left side is $\geq c_-\|u - u_h\|_H^2$, and the right side is $\leq c_+\|u - u_h\|_H\|u - v_h\|_H$. This leads to

$$c_-\|u - u_h\|_H^2 \leq c_+\|u - u_h\|_H\|u - v_h\|_H.$$

Cancelling the factor $\|u - u_h\|_H$ on both sides and recalling that $v_h \in \mathbb{B}_h$ is arbitrary, we obtain the estimate of the lemma.

With the aid of Cea's lemma, the accuracy of spline approximations is easily determined. For example, if the norm of the Hilbert space involves first order partial derivatives, as is the case for $H = H_\Gamma^1$, then the error of the best approximation is usually of order $O(h^n)$, where h is the grid width and n is the degree of the finite elements.

9.2 ▪ Weighted B-Splines

In order to approximate a function u on a domain D, we can simply use uniform spline spaces $\mathbb{B}_h = S_\xi^{(n,\ldots,n)}(D)$ with equal coordinate degrees n and grid width h:

$$\xi_\nu : \xi_{\nu,0}, \xi_{\nu,1} = \xi_{\nu,0} + h, \ldots, \xi_{\nu,m_\nu+n},$$

where the parameter hyperrectangle $D_\xi^{(n,\ldots,n)}$ is a grid conforming bounding box for D. As remarked in Section 7.3, it is convenient to work with a rectangular coefficient array. To this end, we view $u_h = \sum_{k \sim D} u_k b_k \in \mathbb{B}_h$ as the restriction of a spline in $S_\xi^{(n,\ldots,n)}$, i.e.,

$$u_h(x) = \sum_{k_1=0}^{m_1-1} \cdots \sum_{k_d=0}^{m_d-1} u_k b_k(x), \quad x \in D,$$

with irrelevant coefficients ($k \not\sim D$) set to zero.

Perhaps somewhat surprisingly, the simple procedure works well for unconstrained variational problems. Just restricting B-splines to the simulation region provides very accurate finite element approximations u_h for problems with natural boundary conditions. An example is the Neumann problem for Poisson's equation:

$$-\Delta u = f \text{ in } D, \quad \partial_\perp u = 0 \text{ on } \partial D,$$

where f satisfies the (necessary) compatibility condition $\int_D f = 0$. The matrix and the right side of the Ritz–Galerkin system are

$$G : g_{k,k'} = \int_D (\operatorname{grad} b_k)^{\mathrm{t}} \operatorname{grad} b_{k'}, \quad F : f_k = \int_D b_k f,$$

respectively. The solution u is only determined up to an additive constant. Accordingly, the constant vector lies in the kernel of G. This does not present any problem since the compatibility condition implies that $\sum_k f_k = 0$, guaranteeing the solvability of the positive semidefinite system $GU = F$.

To incorporate essential boundary conditions, we resort to an idea already proposed by Kantorowitsch and Krylow and first applied to spline spaces by Schultz.

> **Weighted Basis**
>
> Let w be a function with bounded gradient, which is positive on D and vanishes linearly on a subset Γ of ∂D. Then, the weighted uniform B-splines with grid width h,
>
> $$B_k = w b_k, \quad k \sim D,$$
>
> which have some support in D, span an admissible finite element subspace $\mathbb{B}_h = w S_\xi^{(n,\ldots,n)}(D)$ for problems with homogeneous essential boundary conditions on Γ.

Positivity of the weight function is crucial. If $w(x) = 0$ for some $x \in D$, all approximations u_h vanish at x. Unless the solution u accidentally has a zero at x, this causes a severe loss of accuracy. Similarly, if w vanishes superlinearly on Γ, solutions with nonzero gradient on Γ cannot be well approximated.

☐ **Example:**

The figure illustrates the weighted basis for biquadratic B-splines and a domain D bounded by a circle and an ellipse:

$$\Gamma:\; w(x_1,x_2) = x_1^2 + x_2^2 - 1 = 0, \quad \partial D\backslash\Gamma:\; 1-(x_1-1)^2/9 - x_2^2/4 = 0.$$

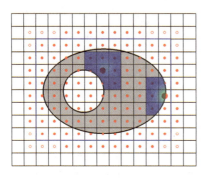
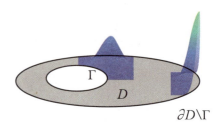

We see that multiplication with the weight function w yields exact fulfillment of essential homogeneous boundary conditions on Γ. Moreover, in the interior of the domain, the qualitative form of the B-splines is hardly affected by w.

As usual, we visualize the free parameters by nodes $\xi_k^{(n,n)}$ (dots) at the centers of the supports of the relevant B-splines b_k, $k \sim D$. Circles mark the centers of the remaining irrelevant B-splines in the basis for $S_\xi^{(n,n)}$. The thick grid lines indicate the boundary of the parameter rectangle $D_\xi^{(n,n)}$, which encloses D. ☐

Weight functions can be constructed in various ways. For many engineering shapes, ad hoc definitions are possible, as in the previous example. A general purpose solution is provided by the smoothed distance function. Moreover, splines yield very flexible approximations to any type of weight functions.

If essential boundary conditions are prescribed along the entire boundary and the domain is defined in terms of simple subsets, Rvachev's R-function method is a convenient choice. It combines elementary weight functions according to Boolean operations and is thus particularly well suited for simulations in conjunction with constructive solid geometry.

R-Functions

Assume that the domains D_ν are described in implicit form by the weight functions w_ν, i.e., $D_\nu: w_\nu > 0$. Then, the standard R-functions

$$r_c(w_1) = -w_1, \qquad r_\backslash(w_1,w_2) = w_1 - w_2 - \sqrt{w_1^2 + w_2^2},$$
$$r_\cup(w_1,w_2) = w_1 + w_2 + \sqrt{w_1^2 + w_2^2}, \qquad r_\cap(w_1,w_2) = w_1 + w_2 - \sqrt{w_1^2 + w_2^2}$$

describe the domains

$$D_1^c,\; D_1\backslash D_2,\; D_1 \cup D_2,\; D_1 \cap D_2,$$

respectively.

9.2. Weighted B-Splines

Let us check, e.g., the formula for $r_\cap$. Since $x \in D_1 \cap D_2 \Leftrightarrow x \in D_1 \wedge x \in D_2$, we have to show that
$$w_1 + w_2 - \sqrt{w_1^2 + w_2^2} > 0 \quad \Leftrightarrow \quad w_1 > 0 \wedge w_2 > 0.$$

Bringing the square root to the right side, the left inequality is equivalent to
$$w_1 + w_2 > 0 \ \wedge \ w_1^2 + 2w_1 w_2 + w_2^2 > w_1^2 + w_2^2.$$

After simplification, the second inequality becomes $w_1 w_2 > 0$ and states that w_1 and w_2 have the same sign. In conjunction with the inequality $w_1 + w_2 > 0$, it follows that the admissible pairs (w_1, w_2) lie in the first quadrant, as claimed.

☐ **Example:**
The figure shows a small sequence of Rvachev operations.

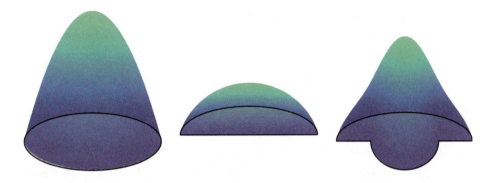

We start with a weight function for a disc:
$$D_1: w_1(x_1, x_2) = 4 - x_1^2 - x_2^2 > 0.$$

Forming the intersection with the half-plane
$$D_2: w_2(x_1, x_2) = x_2 > 0,$$

we obtain a weight function for a half-disc:
$$D_3: w_3 = w_1 + w_2 - \sqrt{w_1^2 + w_2^2} > 0.$$

Finally,
$$w_5 = w_3 + w_4 + \sqrt{w_3^2 + w_4^2}, \quad w_4(x_1, x_2) = 1 - x_1^2 - x_2^2/4,$$

is a weight function for the mushroom shaped domain on the right. ☐

The weighted basis functions $B_k = w b_k$ share all properties of standard finite elements, except for stability. For B-splines b_j, which do not have at least one grid cell of their support in the simulation domain D, the norm of $w b_j$ is very small. For moderate grid width, this does not present any problems. In fact, the weighted basis usually provides adequate approximations. However, for certain algorithms, stability as $h \to 0$ can be crucial. To obtain stable finite elements, we combine neighboring B-splines with support near the boundary, forming so-called weighted extended B-splines (web-splines)

$$\tilde{B}_i = \sum_k e_{i,k}(w b_k), \quad i \in I,$$

as was proposed by Reif, Wipper, and Höllig. The set I comprises all indices of inner B-splines, i.e., those b_i with at least one grid cell of their support in D.

The mathematics leading to the proper choice of the extension coefficients $e_{i,k}$ is somewhat subtle. However, implementing the change $B_k = w b_k \to \tilde{B}_i$ of the finite elements is straightforward. In effect, the amount of work is comparable to a sparse preconditioning procedure, since the (generalized) matrix $(e_{i,k})_{i \in I, k \sim D}$ has only a few off-diagonal entries.

9.3 ▪ Isogeometric Elements

Domains D, arising in engineering simulations, often have simple parametrizations. In such cases, it can be convenient to utilize the parametrization in the construction of finite elements. This is made more precise in the following definition. It describes the basic idea of isogeometric analysis, a finite element technique introduced by Hughes, Cotrell, and Bazilevs in the simplest setting.

> **Isogeometric Elements**
> Assume that
> $$R \ni t \mapsto x = \varphi(t) \in D$$
>
> is a smooth bijective parametrization of a domain D over a hyperrectangle R. Then, the B-splines b_k, which span an appropriate subspace of a spline space $S_\tau^{(n,\dots,n)}$ with parameter hyperrectangle $D_\tau^{(n,\dots,n)} = R$, can be composed with φ to form so-called isogeometric elements
> $$B_k(x) = b_k(\underbrace{\varphi^{-1}(x)}_{t}), \quad k \in K,$$
>
> on D. The knot sequences τ have to be chosen so that essential boundary conditions are satisfied.

□ **Example:**

The construction of biquadratic isogeometric elements is illustrated in the figure. The domain D, shown on the right, is bounded by two parabolas and two line segments and can thus be conveniently parametrized in Bézier form:

$$[0,1]^2 \ni t \mapsto x = \varphi(t) = \sum_{\ell_1=0}^{1} \sum_{\ell_2=0}^{2} c_\ell b_\ell^{(1,2)}(t)$$

$$= ((1-t_1)t_2 + 4t_1 t_2, (1-t_1)(1+t_2^2) + 2t_1 t_2^2),$$

where the Bézier control net C forms two quadrilaterals. On the vertical line segment $\Gamma = \varphi([0,1],0)$ essential boundary conditions are imposed; i.e., the isogeometric elements

9.3. Isogeometric Elements

must vanish there. Accordingly, we choose the knot sequences

$$\tau_{1,0}, \tau_{1,1}, \ldots : -2h, -h, \ldots, 1+2h,$$
$$\tau_{2,0}, \tau_{2,1}, \ldots : -h, 0, 0, h, 2h, \ldots, 1+2h,$$

as shown in the left figure. Only those B-splines which vanish along $\varphi^{-1}(\Gamma) = [0,1] \times \{0\}$ are included in the finite element basis, i.e.,

$$K = \{0, 1, \ldots\} \times \{1, 2, \ldots\}.$$

The B-splines $b^{(2,2)}_{(k_1,0),\tau}$ with $b_k(t_1, 0) \neq 0$ are discarded. In computations we simply set the coefficients of these B-splines b_k, $k \notin K$, to zero. Then, we can work with the full spline space $S^{(n,n)}_\tau$.

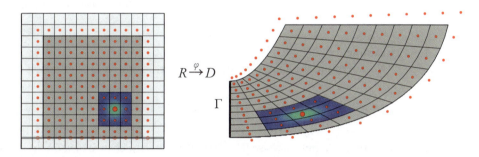

We visualize the B-splines b_k and the isogeometric elements $B_k = b_k \circ \varphi^{-1}$ by nodes (solid dots) in the parameter rectangle R and the domain D. The nodes $\xi^{(n,n)}_k$ for B_k are the images of the multivariate knot averages $\tau^{(n,n)}_k$ under the parametrization φ. For some B-splines, the node $\tau^{(n,n)}_k$ lies outside of the parameter rectangle R. For such k, $\xi^{(n,n)}_k$ can only be displayed if φ can be extended beyond the boundary of R. B-splines, which are not included in the basis, are marked with circles in the parameter rectangle.

The figure also shows the images of the grid lines under the mapping φ. This allows us to assess the distortion caused by the isogeometric transformation. □

The definition of isogeometric elements involves the inverse of the mapping $\varphi : R \to D$. Although it is usually not possible to determine φ^{-1} explicitly, this is not a severe drawback. Via a change of variables, all computations can be performed on the parameter domain R, as is familiar from isoparametric elements.

Finite Element Integrals

Assume that $u(x) = v(t)$, where the variables are related by a smooth bijective parametrization $R \ni t \mapsto x = \varphi(t) \in D$. Then,

$$\int_D f(x, \operatorname{grad}_x u, \ldots) \, dx = \int_R f(\varphi(t), (\varphi'(t)^{-1})^{\mathrm{t}} \operatorname{grad}_t v, \ldots) |\det \varphi'(t)| \, dt,$$

where φ' denotes the Jacobi matrix of φ.

The identity for the finite element integrals follows from the formula for changing variables in multiple integrals:

$$\int_D p(x)\,dx = \int_R q(t)\,|\det\varphi'(t)|\,dt$$

if $D = \varphi(R)$ and $p(x) = q(t)$. It remains to comment on the transformation of the gradient. By the chain rule,

$$(\operatorname{grad}_t \underbrace{u(\varphi(t)))^{\mathrm{t}}}_{v(t)} = (\operatorname{grad}_x u(x))^{\mathrm{t}}\varphi'(t),$$

noting that the gradient is a column vector. Multiplying by $\varphi'(t)^{-1}$ from the right and transposing the identity yield $\operatorname{grad}_x u = (\varphi'(t)^{-1})^{\mathrm{t}}\operatorname{grad}_t v$, as claimed.

□ **Example:**

We determine the entries of the Ritz–Galerkin matrix G for Poisson's bilinear form:

$$g_{k,k'} = \int_D (\operatorname{grad}_x B_k(x))^{\mathrm{t}}\operatorname{grad}_x B_{k'}(x)\,dx.$$

In view of the definition of isogeometric elements, the change of variables $x = \varphi(t), t \in R$, yields

$$\int_R (\operatorname{grad}_t b_k(t))^{\mathrm{t}}\varphi'(t)^{-1}(\varphi'(t)^{-1})^{\mathrm{t}}\operatorname{grad}_t b_{k'}(t)\,|\det\varphi'(t)|\,dt$$

as an expression for the matrix entries. Hence, the computations involve just the ordinary B-splines on the tensor product grid.
□

Of course, using parametrizations φ defined on a single hyperrectangle is very restrictive. Smoothness of φ' and $(\varphi')^{-1}$ implies that the topological structure (corners, edges, etc.) is preserved. Reentrant corners or edges present a problem, too. Therefore, to apply the isogeometric technique, a general domain must be subdivided appropriately, i.e., decomposed into a partition of deformed hyperrectangles:

$$D = \bigcup_\nu D_\nu, \quad D_\nu = \varphi_\nu(R_\nu).$$

In fact, such decompositions may already be provided by CAD descriptions. Generalizing the above concept, we combine the isogeometric elements, corresponding to the subsets D_ν. In other words, we use the basis functions

$$x \mapsto B_{k,\nu}(x) = b_{k,\nu}(\varphi_\nu^{-1}(x)), \quad x \in D_\nu.$$

Clearly, to ensure continuity, consistency at patch boundaries is crucial. This means that B-splines restricted to a common patch boundary must coincide and share the same coefficient.

9.4. Implementation

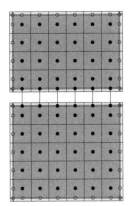
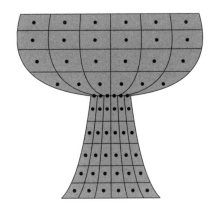

As for a single parameter rectangle, we visualize the degrees of freedom by the nodes $\tau_k^{(n,\ldots,n)}$ and $\xi_k^{(n,\ldots,n)} = \varphi\left(\tau_k^{(n,\ldots,n)}\right)$ (dots). As illustrated in the figure for two dimensions, in view of consistency, nodes $\xi_{k_1,k_2}^{(n,n)}$ on a common patch boundary are shared by the corresponding B-splines of the neighboring patches. Moreover, if essential boundary conditions are imposed as in the figure, coefficients associated with nodes on outer boundaries are set to zero (circles).

Some of the elegance of the isogeometric approach is lost when an irregular patch structure is necessary to model complicated domains. In particular, it is not straightforward to achieve more smoothness than continuity across patch boundaries. Moreover, especially for three-dimensional domains, a partition into deformed cubes D_ν is not easy to construct if it is not yet provided by a solid model.

9.4 ▪ Implementation

Implementing B-spline based finite element methods is extremely simple. For weighted B-splines $B_k = w b_k$, $k \sim D$, as well as for isogeometric elements $B_k = b_k \circ \varphi^{-1}$, $k \in K$, parametrized over a single hyperrectangle, the numerical approximation can be represented in the form

$$u_h = \sum_{k_1=0}^{m_1-1} \cdots \sum_{k_d=0}^{m_d-1} u_k B_k,$$

where b_k are the B-splines spanning a spline space $S_\xi^{(n,\ldots,n)}$ with parameter hyperrectangle $R = D_\xi^{(n,\ldots,n)}$ and irrelevant coefficients u_k set to zero. Hence, the parameters or unknowns form a d-dimensional array of dimension $m_1 \times \cdots \times m_d$, which we denote by

$$U = (u_k)_{k \sim R}.$$

We now turn to the assembly of the Ritz–Galerkin system

$$GU = F, \quad g_{k,k'} = a(B_k, B_{k'}), \quad f_k = \lambda(B_k).$$

Usually, the bilinear form $a(\cdot,\cdot)$ and the linear functional λ are defined in terms of integrals over subsets of R; Poisson's problem may serve as an elementary example. Hence, the entries of G and F can be computed by summing the contributions from each grid cell

$$D_\ell = [\xi_{\ell_1}^1, \xi_{\ell_1+1}^1] \times \cdots \times [\xi_{\ell_d}^d, \xi_{\ell_d+1}^d].$$

If we denote these contributions by $a_{k,k',\ell}$ and $\lambda_{k,\ell}$, respectively, then

$$g_{k,k'} = \sum_\ell a_{k,k',\ell}, \quad f_k = \sum_\ell \lambda_{k,\ell}.$$

Clearly, the integrals are nonzero only if the support of the B-spline b_k involved overlaps D_ℓ, i.e., if

$$k \sim \ell \quad \Leftrightarrow \quad \ell_\nu - n \leq k_\nu \leq \ell_\nu, \nu = 1,\dots,d.$$

For the Ritz–Galerkin matrix this implies that each row (and column) has at most $(2n+1)^d$ nonzero entries; b_k and $b_{k'}$ have a common grid cell only if $|k_\nu - k'_\nu| \leq n$ for $\nu = 1,\dots,d$. It is, therefore, convenient to refer to the nonzero entries $g_{k,k'}$ by the index k and the offset $s = k' - k \in \{-n,\dots,n\}^d$. Accordingly, we store the relevant part of G in an array $\tilde{G} = (\tilde{g}_{k_1,\dots,k_d,s_1,\dots,s_d})_{k \sim R, |s_\nu| \leq n}$ of dimension

$$m_1 \times \cdots \times m_d \times (2n+1) \times \cdots \times (2n+1).$$

Similarly, the right side F is stored in an array $\tilde{F}$ of dimension $m_1 \times \cdots \times m_d$.

With these conventions, the Ritz–Galerkin system can be assembled via the following algorithm.

Assembly of the Ritz–Galerkin System

The entries of the matrix and right side of the Ritz–Galerkin system are computed by adding the contributions from each grid cell:

$$\tilde{G} = 0, \tilde{F} = 0$$
$\text{for } D_\ell \subseteq R$
$\quad \text{for } k \sim \ell$
$\qquad \tilde{f}_k = \tilde{f}_k + \lambda_{k,\ell}$
$\qquad \text{for } k' \sim \ell$
$\qquad\qquad \tilde{g}_{k,k'-k} = \tilde{g}_{k,k'-k} + a_{k,k',\ell}$
$\qquad \text{end}$
$\quad \text{end}$
end

This looks extremely simple, and it is—no index lists, just loops over multiple arrays. We have only to supplement routines for computing the integrals $a_{\ell,k,k'}$ and $\lambda_{\ell,k}$. For isogeometric elements, this is straightforward. The domains of integration are the entire grid cells D_ℓ for all ℓ. Hence, standard tensor product Gauß formulas can be used. For weighted B-splines, this is possible only for inner grid cells. If the boundary of D intersects D_ℓ, integration is more complicated. As illustrated in the figure, we have to decompose $D \cap D_\ell$ into subsets which can be mapped to standard domains in order to preserve accuracy. The corresponding nodes $\eta_{\ell,\alpha}$ (marked with stars in the figures) and weights $\omega_{\ell,\alpha}$ are best computed in a preprocessing step. Then, the simplicity of the assembly algorithm

9.4. Implementation

is preserved. For all integrands ψ, the numerical approximation is of the form

$$\int_{D \cap D_\ell} \psi \approx \sum_\alpha \omega_{\ell,\alpha} \psi(\eta_{\ell,\alpha}),$$

regardless of whether $D_\ell \subseteq D$ or not.

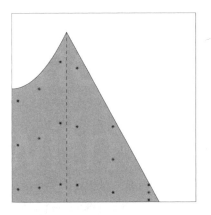
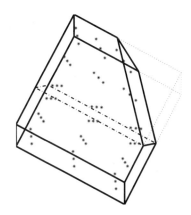

A small adjustment of the assembled system is necessary. Not all B-splines are relevant. For weighted bases, some B-splines b_j have support outside the simulation domain. Hence, the corresponding rows and columns in the Ritz–Galerkin matrix have zero entries. For isogeometric elements, some boundary B-splines $b_j \circ \varphi^{-1}$ do not belong to the basis due to essential boundary conditions. In both cases, we define

$$\tilde{f}_j = 0, \quad \tilde{g}_{j,j} = 1, \quad \tilde{g}_{j,k} = \tilde{g}_{k,j} = 0, k \neq j,$$

for such j and all k. As a result, the irrelevant coefficients u_j of the Ritz–Galerkin approximation $u_h = \sum_k u_k B_k$ will be zero.

☐ **Example:**

We illustrate the assembly algorithm for a trivial, yet instructive, univariate example with

$$a(u,v) = \int_0^1 u'v' + 3uv, \quad \lambda(u) = \int_0^1 2u$$

and no (essential) boundary conditions ($\Gamma = \emptyset$).

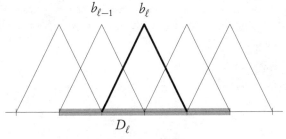

Using linear B-splines with grid width h and knot sequence

$$\tau_0 = -h, 0, \ldots, 1+h = \tau_{m+1},$$

as shown in the figure, the matrix G is tridiagonal, and $\tilde{G}$ has dimension $m \times 3$, $h = 1/(m-1)$. The interval $\overline{D} = [0,1]$ is partitioned into $m-1$ grid cells $D_\ell = [\ell-1,\ell]h$. Hence, we obtain the following ranges for the loops in the assembly algorithm:

$$\begin{aligned}
&\textbf{for } D_\ell \subseteq R: && \ell = 1,\dots,m-1, \\
&\textbf{for } k \sim \ell: && k = \ell-1,\ell, \\
&\textbf{for } k' \sim \ell: && k' = \ell-1,\ell\ .
\end{aligned}$$

The contributions of the linear functional and the bilinear form corresponding to the ℓth grid cell can be determined explicitly:

$$\lambda_{k,\ell} = \int_{(\ell-1)h}^{\ell h} 2 b_k = h,$$

$$a_{k,k',\ell} = \int_{(\ell-1)h}^{\ell h} b_k' b_{k'}' + 3 b_k b_{k'} = \begin{cases} \frac{1}{h} + h & \text{for } k = k', \\ -\frac{1}{h} + \frac{h}{2} & \text{for } |k - k'| = 1, \end{cases}$$

for $k,k' \sim \ell$. As a concrete example, we obtain for $h = 1/4$

$$\tilde{F} = \begin{pmatrix} 0.25 \\ 0.5 \\ 0.5 \\ 0.5 \\ 0.25 \end{pmatrix}, \quad \tilde{G} = \begin{pmatrix} 0 & 4.25 & -3.875 \\ -3.875 & 8.5 & -3.875 \\ -3.875 & 8.5 & -3.875 \\ -3.875 & 8.5 & -3.875 \\ -3.875 & 4.25 & 0 \end{pmatrix}$$

as a result of the assembly algorithm.

The first and last row of $\tilde{F}$ and $\tilde{G}$ differ from the standard pattern since the grid cells $[-0.25,0]$ and $[1,1.25]$ do not contribute to the matrix entries. For the same reason, the zeros in the upper left and lower right positions do not correspond to entries of the tridiagonal matrix G. $\qquad\square$

The storage scheme is also well suited for the iterative solvers of the Ritz–Galerkin equations. For most such schemes we have to implement a matrix-vector multiplication. This can be done very efficiently.

Multiplication by the Ritz–Galerkin matrix

Assume that the matrix $(g_{k,k'})_{k,k'\sim R}$ is stored in an array $(\tilde{g}_{k,s})_{k\sim R,|s_\nu|\leq n}$ with the second index s corresponding to the offsets $k' - k$. Then, for a vector $(u_k)_{k\sim R}$, the product $V = GU$ can be computed with the following algorithm:

$$\begin{aligned}
&V = 0 \\
&\textbf{for } s \in \{-n,\dots,n\}^d \\
&\quad \textbf{for } k \sim R \\
&\qquad\quad v_k = v_k + \tilde{g}_{k,s} u_{k+s} \\
&\quad \textbf{end} \\
&\textbf{end}
\end{aligned}$$

where entries u_{k+s} with indices $\not\sim R$ are set to zero.

□ Example:

We consider again the model problem of the previous example with a tridiagonal Ritz–Galerkin matrix G, stored in an $m \times 3$ array

$$\tilde{G} = \begin{pmatrix} 0 & g_{0,0} & g_{0,1} \\ g_{1,0} & g_{1,1} & g_{1,2} \\ & \vdots & \\ g_{m-2,m-3} & g_{m-2,m-2} & g_{m-2,m-1} \\ g_{m-1,m-2} & g_{m-1,m-1} & 0 \end{pmatrix}.$$

To implement the matrix multiplication, we pad the input vector with zeros,

$$\tilde{U} = \begin{pmatrix} 0 & u_0 & \cdots & u_{m-1} & 0 \end{pmatrix}^{\mathrm{t}}.$$

Then $V = GU$ is computed by

$$\tilde{G}(:,-1).*\tilde{U}(-1:m-2) + \tilde{G}(:,0).*\tilde{U}(0:m-1) + \tilde{G}(:,1).*\tilde{U}(1:m),$$

where $.*$ denotes pointwise multiplication. The general case of a $2d$-dimensional matrix and a d-dimensional vector is completely analogous. The inner loop involves a pointwise multiplication of sections of $\tilde{G}$ with shifts of a padded vector U. □

9.5 ▪ Applications

As a first very simple example we consider the flow of an incompressible fluid in a channel with varying cross section as depicted in the figure. By conservation of mass, the velocity field V satisfies

$$\operatorname{div} V(x) = 0, \quad (x_1, x_2) \in D.$$

Since the domain D representing the channel is simply connected, there exists a potential u, i.e.,

$$V = -\operatorname{grad} u, \quad -\Delta u = 0.$$

At the vertical boundaries Γ_{left} and Γ_{right}, the normal velocity $-\partial_\perp u$ is assumed to be constant and equal to s and $-s$, respectively. Moreover, $\partial_\perp u = 0$ on the curved boundaries. With the calculus of variations it is not difficult to show that a solution u of this boundary value problem can be characterized as a minimizer of the quadratic functional

$$u \mapsto Q(u) = \frac{1}{2} \int_D |\operatorname{grad} u|^2 - s \int_{\Gamma_{\text{left}}} u + s \int_{\Gamma_{\text{right}}} u$$

over all u in the Sobolev space $H^1(D)$.

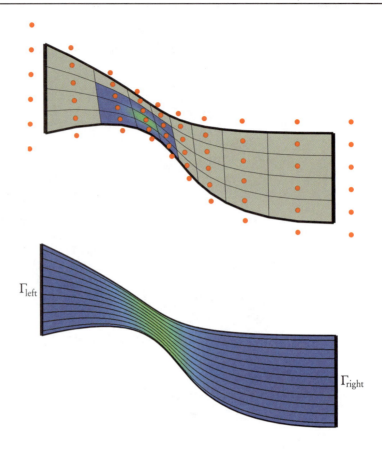

The flow problem serves as an ideal example for the isogeometric method. We represent the channel D in Bézier form, using a mixed cubic/linear parametrization:

$$[0,2] \times [0,1] = R \ni t \mapsto x = \varphi(t) = \sum_{\ell_1=0}^{3} \sum_{\ell_2=0}^{1} c_\ell b_\ell^{(3,1)}(t_1/2, t_2),$$

with $b_\ell^{(3,1)}(s) = \binom{3}{\ell_1}(1-s_1)^{3-\ell_1} s_1^{\ell_1} (1-s_2)^{1-\ell_2} s_2^{\ell_2}$ and control points

$$\begin{array}{rcccccl}
 & & (0,20) & (18,12) & (9,10) & (33,10) & = c_{3,1}, \\
c_{0,0} = & & (0,10) & (21,15) & (9,0) & (33,0). &
\end{array}$$

We choose biquadratic isogeometric elements with equally spaced knots and grid width h in both coordinate directions:

$$\tau_1: -2h, -h, \ldots, 2+2h, \quad \tau_2: -2h, -h, \ldots, 1+2h.$$

Since no essential boundary conditions are prescribed, all B-splines are relevant, leading to the isogeometric basis

$$B_k = b_{k,\tau}^{(2,2)} \circ \varphi^{-1}, \quad k \in K = \{0,1,\ldots,2/h+1\} \times \{0,1,\ldots,1/h+1\},$$

visualized by dots in the top figure which also shows the image of the regular grid. The bottom figure shows the stream lines and magnitude of the velocity.

9.5. Applications

As expected, higher degree n leads to substantially more accurate approximations. On the top, the figure shows the logarithmic (base 10) errors in the L_2- and the H^1-norm up to degree $n = 5$ for a sequence of grids. On the bottom, the error reduction rates (negative logarithm, base 2) are visualized.

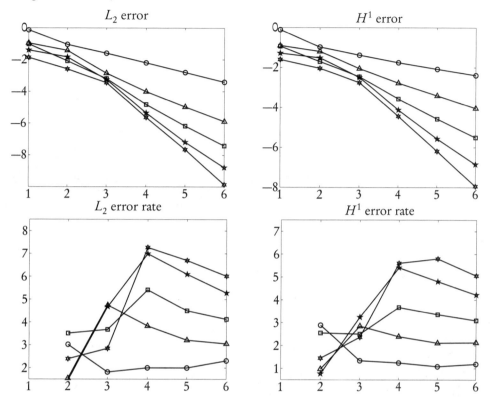

In accordance with the remark after the proof of Cea's lemma, halving the grid width leads to an error reduction by roughly a factor 2^{-n} (bottom right).

The second example concerns one of the key applications of finite elements: linear elasticity. A typical situation is shown in the figure. An elastic solid occupying a volume D is fixed at a portion Γ of its boundary and subject to an internal force with density f. This results in a small deformation:

$$D \ni (x_1, x_2, x_3) \mapsto x + (u_1(x), u_2(x), u_3(x))$$

with the displacement u visualized by an arrow in the figure.

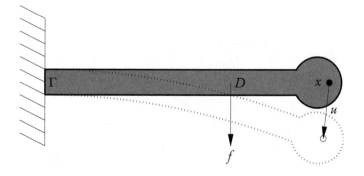

We describe the mathematical model without derivation.

Linear Elasticity

The displacement (u_1, u_2, u_3) caused by an elastic deformation minimizes the quadratic energy functional

$$\mathcal{Q}(u) = \frac{1}{2} \int_D \sum_{\nu,\nu'=1}^{3} \varepsilon_{\nu,\nu'}(u) \sigma_{\nu,\nu'}(u) - \int_D \sum_{\nu=1}^{3} f_\nu u_\nu, \quad u_\nu \in H^1_\Gamma(D),$$

where ε is the strain and σ is the stress tensor, defined by

$$2\varepsilon_{k,\ell} = \partial_k u_\ell + \partial_\ell u_k, \quad \sigma_{k,\ell} = \lambda \operatorname{trace} \varepsilon \, \delta_{k,\ell} + 2\mu \varepsilon_{k,\ell},$$

respectively. The constants λ and μ are the Lamé coefficients, which describe the elastic properties of the material.

Via the calculus of variations, the displacement can also be characterized by the Lamé–Navier boundary value problem

$$-\operatorname{div} \sigma(u) = f \text{ in } D, \quad u = 0 \text{ on } \Gamma, \quad \sigma(u)\eta = 0 \text{ on } \partial D \setminus \Gamma,$$

with η the outward unit normal of ∂D.

To define a finite element approximation u_h, we use weighted splines for each component, i.e.,

$$(u_h)_\nu \in w S_\xi^{(n,n,n)}(D), \quad \nu = 1, 2, 3.$$

The knot sequences ξ_ν are uniform and $D \subseteq D_\xi^{(n,n,n)}$. The weight function

$$(x_1, x_2, x_3) \to w(x)$$

incorporates the essential boundary conditions; i.e., it vanishes linearly on Γ. As described in Sections 9.2 and 9.4, it is convenient to work with rectangular coefficient arrays. Accordingly, the finite element approximation has the form

$$u_h = \sum_{k_1=0}^{m_1-1} \sum_{k_2=0}^{m_2-1} \sum_{k_3=0}^{m_3-1} \begin{pmatrix} u_{k,1} \\ u_{k,2} \\ u_{k,3} \end{pmatrix} w b_k$$

with irrelevant coefficients set to zero.

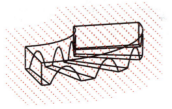
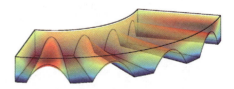

9.5. Applications

The figure shows a simplified shape of a concrete bridge ($\lambda = 8.3$ kN/mm^2, $\mu = 12.5$ kN/mm^2). For this example, the elementary weight function

$$(x_1, x_2, x_3) \mapsto w(x) = x_3$$

can be used. For a constant vertical force, representing gravity, we obtain a fairly accurate approximation already with relatively few parameters. For example, for a basis with 3×1875 tricubic B-splines indicated by solid nodes, the relative error of the displacement in the L_2-norm is less than $2 \cdot 10^{-4}$.

If essential boundary conditions are specified on a planar boundary portion $\Gamma \subset \{x \in \mathbb{R}^3 : x_3 = 0\}$, as in the previous example, a weight function is not necessary. Instead, we can use the knot sequence

$$\xi_3 : -h, \underbrace{0, \ldots, 0}_{n \text{ times}}, h, 2h, \ldots$$

in the direction perpendicular to Γ and discard the B-splines $b_{(k_1, k_2, 0)}$, which are nonzero on Γ. In this way the boundary condition $u = 0$ is incorporated directly into the B-spline basis. However, we cannot merely use translates of a single B-spline. Therefore, also in this simple situation, it seems that the weighted method is slightly more efficient.

Notes and Comments

Below, we list references and comment on the history of the concepts and results presented in this book. Especially for basic topics, many good sources exist. We have cited what appear to be historically the first articles and those that best fit our approach.

Chapter 1. The basic facts about approximation with polynomials in Sections 1.1, 1.2, 1.3, and 1.6 can be found in almost any text on approximation and numerical methods; cf., e.g., [32, 41, 74, 116]. However, it is certainly illuminating to consult the classical papers by Taylor [118], Waring [121], and Hermite [62]. The iterative construction of polynomial interpolants was independently discovered by Aitken [2] and Neville [85]. Perhaps somewhat unexpected, the analysis of the 4-point scheme and related interpolatory subdivision techniques is quite subtle. The article by Dubuc [44] is the first in a series of interesting papers (cf., e.g., [45, 37, 54]). The example illustrating the divergence of polynomial interpolants is due to Runge [102]. Bernstein polynomials were probably first used by Bernstein [6] in his proof of the Weierstrass approximation theorem [122]. For a comprehensive discussion of their properties, also including basic interpolation methods, see [49, 80].

Chapter 2. Bézier curves were independently introduced by de Casteljau [25, 26] and Bézier [7, 8, 9, 10]. For an excellent description of their geometric and algorithmic properties, we refer the reader to the classical book by Farin [49] and the references cited therein. The famous de Casteljau algorithm [26] is the basic example of many similar fundamental triangular schemes in approximation and design. It not only evaluates but also subdivides Bézier curves, as was rigorously proved in [113]. The cubic geometric Hermite interpolant proposed by de Boor, Höllig, and Sabin [23] was the first scheme with a nonstandard order of convergence. The phenomenon of superconvergence was subsequently investigated by many authors (cf., e.g., [91, 104, 60, 72] as well as [68] for an interesting general conjecture). Highly accurate approximation methods rely on concepts from differential geometry (curvature, geometric smoothness, etc.), which, in the context of CAGD, are beautifully described by Boehm [12] (cf. also the book by Boehm and Prautzsch [13]) and in the fundamental work of Barsky and DeRose [4, 39, 5].

Chapter 3. The use of rational parametrizations dates back to Coons [33] and Forrest [55]. The generalization of algorithms and properties for polynomial curves is straightforward and is described, e.g., by Farin in [47, 50, 51]. In the first cited paper, Farin introduced "weight points," which remove the redundancy among the weights in an elegant fashion. For evaluation and differentiation, the use of homogeneous control points allows the direct application of polynomial techniques; cf., e.g., the book by Hoschek and Lasser [70]. The important special case of conic sections is a classical topic. Among the vast literature, we refer the reader to the article by Lee [79] for a discussion from a modeling viewpoint.

Chapter 4. The canonical reference for the basic B-spline calculus, pertaining to approximation schemes as well as algorithms, is de Boor's *A Practical Guide to Splines* [19]; cf. also [17, 18]. We use a slightly different approach, defining B-splines via their recurrence relations [15, 35], as was done by de Boor and Höllig in [21]. Marsden's identity [83] is then the crucial tool for deriving analytical properties and identities for spline spaces. De Boor's algorithms for evaluation and differentiation of splines follow directly from the fundamental recurrence relations. For uniform splines, as studied extensively by Schoenberg [107], formulas become particularly elegant.

Chapter 5. Schoenberg's scheme appeared first in [105] (cf. [78] for an elegant proof of its shape-preserving properties). A systematic theory of quasi-interpolants providing optimal error estimates was developed by de Boor and Fix [14, 20] and Lyche and Schumaker [82, 111] in considerably more generality than presented in our book (cf. also the book by Strang and Fix [117] for results in the context of finite elements and the article by Reif [94] for an interesting relation to orthogonal projections). The famous stability result in Section 5.4 is due to de Boor [14] and was proved with the aid of dual functionals inherent in any local spline projection (cf., however, [64] for an argument based on knot insertion). Our proof of the Schoenberg–Whitney theorem [109] follows de Boor [16], who obtained more general results proving also the total positivity of the interpolation matrix. The characterization of the natural spline interpolant was obtained by Holladay [63], and the smoothing spline is due to Schoenberg and Reinsch [106, 98].

Chapter 6. Spline curves were introduced to geometric modeling in a systematic fashion by Gordon and Riesenfeld [99, 59], who established basic properties of the control net representation. For the generalization to rational parametrizations, see, e.g., [120, 47, 119], and [87] for a comprehensive description of NURBS techniques. The precise estimate for the distance to the control polygon is due to Reif [96]. Fundamental for modeling and design techniques are the geometric knot insertion and subdivision strategies, obtained by Boehm [11] and Cohen, Lyche, and Riesenfeld [30]; cf. also [100] for the famous special case of Chaikin's algorithm [29]. While knot insertion for spline functions was already part of de Boor's B-spline package [18], the geometric interpretation for B-spline curves is truly beautiful. Uniform subdivision with its striking simplicity is among our favorite algorithms. An immediate application (see Lane and Riesenfeld [77]) is the variation diminution property, originally discovered by Schoenberg and Greville [108] in the context of approximation. Evaluation and differentiation do not present anything substantially new for the parametric case; yet, the geometric interpretation of the algorithms is rather pretty. To the authors, the blossoming principle of Ramshaw [92] for conversion to Bézier form is particularly appealing. Spline interpolation techniques were applied to curves by numerous authors; see [49, 70, 31] for some of the available techniques. Historically, Ferguson's report [53] appears to be the first source.

Chapter 7. There are a number of different generalizations of univariate splines. Besides tensor products, as discussed in our book, splines over triangulations [76] and box-splines [22] are probably the most intensively studied variants. The short summary of properties of polynomials of coordinate degree is routine; cf. [49] for a discussion of tensor products of Bernstein polynomials. While the error estimate for multivariate polynomials has been known for a long time (cf., e.g., [40] for a much more general result), the rather elegant proof, presented in our book in a special case, is fairly recent and due to Reif [97]. The definition of tensor product spline spaces is straightforward (cf., e.g., [111]). Likewise, applying the tensor product formalism to algorithms (evaluation, differentiation, etc.) and approximation methods (interpolation and quasi-interpolation) proceeds in a standard fashion and has been described by de Boor [19]; see also [49, 70]. For

Notes and Comments

general error estimates for multivariate splines, the reader can consult, e.g., the textbook by Schumaker [111], as well as the articles [40, 38]; our book covers only a special case, which is adequate for the principal applications. Because of the regular grid of tensor product splines, hierarchical bases are crucial for adaptive refinement. As a consequence, a number of techniques have been proposed by different authors; cf., e.g., [56, 75, 112] for a few examples illustrating the diversity of methods available. Our definition is a variant of the more general hierarchical bases in [65] and closely related to the recent approach by Lyche [81] and Dokken, Lyche, and Petterson [42] via knot insertion.

Chapter 8. Among the many possible references for tensor product Bézier surfaces, the book by Prautzsch, Boehm, and Paluszny [90] also contains a detailed discussion of the conditions for smooth patch connections. The elegant idea of using multiple control points at irregular vertices to achieve tangent plane continuity with the aid of the most elementary C^1-constraints is due to Reif [93]. An equally simple construction for smoothness of arbitrary order seems out of reach—perhaps not even possible. To date, the constructions of Prautzsch [89] and Reif [95] provide the best solutions. For rectangular parameter domains, the topologically trivial case, spline surfaces have become a standard in geometric modeling; cf., e.g., [87] where the more general rational NURBS representation is described in detail. Subdivision surfaces provide an alternative to pure B-spline representations and are comprehensively discussed in the book by Peters and Reif [86]. We describe the famous classical construction by Catmull and Clark [27], which is to date one of the most commonly used schemes; cf. [43] for a first theoretical analysis. For the quadrilateral blending techniques, we refer the reader to Coons's fundamental paper [33] and the article by Gordon [58]. Many variants have been proposed; a survey of some of the existing techniques, also including the challenging topics of n-sided patches and approximations with higher order smoothness, can be found in [70]. Spline solids are just very briefly introduced in our book. We refer the reader to [61, 31] for a more detailed discussion.

Chapter 9. The basic variational principle for characterizing solutions of elliptic boundary value problems is due to Ritz [101] and Galerkin [57]; Céa [28] established the fundamental error estimate. The classical book by Strang and Fix [117] provides an excellent introduction to these topics and all essential aspects of finite element theory. The systematic use of B-splines as trial functions began with the construction of web-splines by Höllig, Reif, and Wipper [69]. This new type of finite element incorporates essential boundary conditions via weight functions, an idea of Kantorowitsch and Krylow [73] which was first applied to splines by Schultz [110]. Weighted approximations have been intensively studied by Rvachev and his collaborators (cf., e.g., the survey [103]), who developed in particular the R-function calculus. However, stability problems were not resolved. By contrast, the web-spline approach yields weighted B-spline bases, which share all basic properties of standard mesh-based finite elements. Another B-spline based technique was introduced by Hughes, Cottrell, and Bazilevs [71]. Their new isogeometric elements use NURBS parametrizations to describe the simulation regions, conveniently utilizing CAD representations. In our book, we give just a very brief outline of the main ideas of both concepts, as was done in [67]. For a comprehensive treatment which includes also algorithmic aspects and typical applications, we refer the reader to the books [65, 34]. Moreover, the sample MATLAB programs for the weighted techniques [66] provide further illustrations.

Appendix

The book does not require advanced mathematical tools. A solid knowledge of basic linear algebra and numerical analysis and of elementary geometry suffices. For Chapter 9 some experience with conventional finite element methods is helpful (cf, e.g., the book by Strang and Fix [117]).

Key definitions and results are listed below for convenience of the reader. They can be found in any of the standard textbooks, e.g., [114, 115, 32, 74, 13, 117].

Basis. A subset $\{b_1, \ldots, b_n\}$ of a finite dimensional vector space V is a basis iff any element $v \in V$ has a unique representation

$$v = \sum_{k=1}^{n} c_k\, b_k$$

for some scalars c_k.

The number n of basis elements is uniquely determined and called the dimension of V.

Bilinear form. A bilinear form $a(\cdot, \cdot) : V \times V \to \mathbb{R}$ on a (real) vector space V is linear in each variable, i.e.,

$$a(\alpha_1 u_1 + \alpha_2 u_2, \beta_1 v_1 + \beta_2 v_2) = \sum_{\nu=1}^{2} \sum_{\mu=1}^{2} \alpha_\nu \beta_\mu a(u_\nu, v_\mu)$$

for all $\alpha_\nu, \beta_\mu \in \mathbb{R}$ and $u_\nu, v_\mu \in V$.

Cauchy–Schwarz inequality. For a norm induced by a scalar product,

$$|\langle u, v \rangle| \leq \|u\|\|v\|.$$

Equality holds iff u and v are linearly dependent.

Conic section. The classical quadratic curves (ellipses, parabolas, and hyperbolas) arise as intersections of a cone with a plane. With respect to a cartesian coordinate system of the plane, they have the implicit representation

$$Q: \quad x^t A x + 2 b^t x + c = 0,$$

where $x = (x_1, x_2)^t$, A is a symmetric matrix, b is a vector, and c is a constant. This equation also includes degenerate cases: a pair of lines, a point, and the empty set.

In homogeneous coordinates $z = (x\lambda \,|\, \lambda) \in \mathbb{R}^3$, an equivalent description is

$$Q: \quad z^t \tilde{A} z = 0,$$

where
$$\tilde{A} = \left(\begin{array}{c|c} A & b \\ \hline b^t & c \end{array} \right)$$
is a symmetric 3×3 matrix.

Convex combination. A convex combination c of $a_k \in \mathbb{R}^d$ is a linear combination with nonnegative coefficients which sum to 1:
$$c = \sum_{k=0}^{n} \alpha_k a_k, \quad \alpha_k \geq 0, \quad \sum_{k=0}^{n} \alpha_k = 1.$$

The set of all such convex combinations is called the convex hull of a_k, $k = 0, \ldots, n$, and denoted by $[a_0, \ldots, a_n]$. In particular, $[a_0, a_1]$ is the line segment connecting the points a_0 and a_1, and $[a_0, a_1, a_2]$ is the triangle with vertices a_0, a_1, a_2.

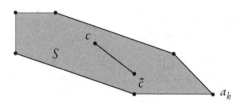

Geometrically, the convex hull is the smallest set S containing $\{a_0, \ldots, a_n\}$ for which
$$c, \tilde{c} \in S \Rightarrow [c, \tilde{c}] \in S.$$

Cross product. For two vectors $u, v \in \mathbb{R}^3$, the cross product is defined as
$$u \times v = \left(\begin{array}{c} u_2 v_3 - u_3 v_2 \\ u_3 v_1 - u_1 v_3 \\ u_1 v_2 - u_2 v_1 \end{array} \right).$$

The vector $u \times v$ is orthogonal to both, u and v, and its length $|u \times v|$ equals the area of the parallelogram spanned by u and v.

Curvature. The (first) curvature $\varkappa(t)$ at a point $p(t)$ on a curve, parametrized by p, is the reciprocal value of the radius of the osculating circle.

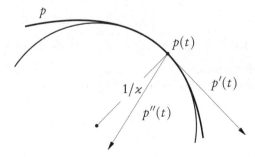

If p is a parametrization with respect to arclength, i.e., if $|p'(t)| = 1$ for all t, then $\varkappa = |p''|$. For an arbitrary regular parametrization,
$$\varkappa = \text{area}(p', p'')/|p'|^3,$$

where area(a, b) denotes the area of the parallelogram spanned by the two vectors a and b and $|a|$ the length of the vector a.

In two dimensions, area$(p', p'') = |\det(p', p'')|$. If the absolute value is omitted, we speak of the signed curvature.

In three dimensions, area$(p', p'') = |p' \times p''|$, where $\times$ denotes the cross product.

Diagonally dominant matrix. A square matrix A is diagonally dominant if

$$|a_{j,j}| > \sum_{k \neq j} |a_{j,k}|$$

for all row indices j.

Divided difference. The divided difference Δ of a function f at the points $x_0, \ldots, x_n$ satisfies the recursion

$$(x_n - x_0)\Delta(x_0, \ldots, x_n)f = \Delta(x_1, \ldots, x_n)f - \Delta(x_0, \ldots, x_{n-1})f.$$

With this identity, $\Delta(\ldots)f$ can be computed in a triangular scheme starting with

$$\Delta(\underbrace{x, \ldots, x}_{(k+1)\text{ times}})f = \frac{f^{(k)}(x)}{k!}.$$

For a smooth function f,

$$\Delta(x_0, \ldots, x_n)f = \frac{f^{(n)}(\xi)}{n!}$$

with $\xi \in [\min x_k, \max x_k]$, which generalizes the mean value theorem.

Hilbert space. A Hilbert space H is a complete vector space with a norm induced by a scalar product, i.e.,

$$\|v\| = \sqrt{\langle v, v \rangle}$$

for $v \in H$.

Homogeneous coordinates. Cartesian coordinates $x = (x_1, \ldots, x_d) \in \mathbb{R}^d$ can be identified with homogeneous coordinates

$$z = (x_1 \lambda, \ldots, x_d \lambda | \lambda) \in \mathbb{R}^{d+1}, \quad \lambda \neq 0,$$

i.e., with lines through the origin.

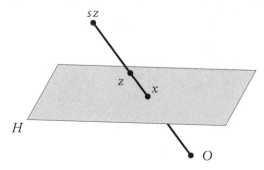

As is illustrated in the figure, the map $z \mapsto x$ is a central projection onto the plane $H : \lambda = 1$. As a consequence, homogeneous coordinates do not distinguish between scalar multiples, $z \sim sz$, which is convenient for certain applications.

Kronecker symbol. The Kronecker symbol δ is defined by

$$\delta_{k,\ell} = \begin{cases} 1 & \text{for } k = \ell, \\ 0 & \text{otherwise}. \end{cases}$$

It is often used in connection with orthogonal expansions. For example, $\sum_k f_k \delta_{k,\ell} = f_\ell$.

Landau symbols. The Landau symbols O and o are used to compare the behavior of two functions f and g in the neighborhood of a point a. We say that

$$f(x) = O(g(x)) \quad \text{for } x \to a$$

if

$$|f(x)| \le c\, g(x), \quad |x - a| < \delta,$$

for some $\delta > 0$ and with a constant c which does not depend on x. The notation $f(x) = o(g(x))$ is used if

$$\lim_{x \to a} \frac{f(x)}{g(x)} = 0,$$

i.e., if $f(x)$ is substantially smaller than $g(x)$ for $x \to a$.

In an analogous fashion, we can use Landau symbols to describe the behavior of functions for $x \to \pm\infty$, or if the variable is a discrete parameter $k \in \mathbb{Z}$.

Least squares line. A linear function $x \to p(x) = u + vx$, which best fits the data (x_k, y_k), $k = 1, \ldots, n$, can be determined by minimizing the mean square error

$$e = \sum_k (y_k - p(x_k))^2.$$

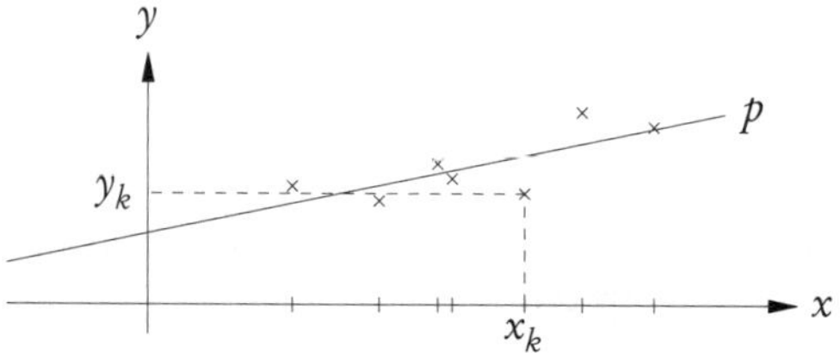

The parameters u and v can be determined by solving the linear equations $\partial_u e = 0 = \partial_v e$.

Linear interpolation. The line segment $[a, b]$, connecting a and b, can be parametrized by

$$p(t) = (1 - t)a + tb, \quad 0 \le t \le 1.$$

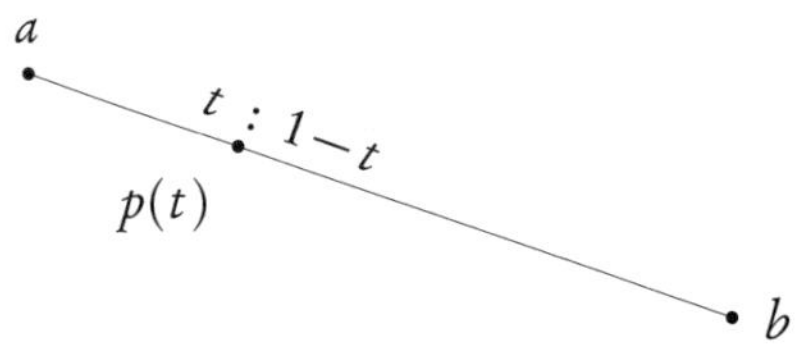

As is illustrated in the figure, $p(t)$ divides $[a, b]$ in the ratio $t : (1 - t)$.

Linear functional. A linear functional $\lambda: V \to \mathbb{R}$ on a (real) vector space V satisfies

$$\lambda\left(\sum_k r_k v_k\right) = \sum_k r_k \lambda(v_k)$$

for all $r_k \in \mathbb{R}$ and $v_k \in V$. It is continuous if

$$\|\lambda\| = \sup_{\|v\|=1} |\lambda(v)|$$

is finite.

Maximum norm. The maximum norm of a vector x is defined as

$$\|x\|_\infty = \max_k |x_k|,$$

and

$$\|A\|_\infty = \max_j \sum_k |a_{j,k}| = \max_{\|x\|_\infty=1} \|Ax\|_\infty$$

is the associated norm of a matrix A.

For a function f,

$$\|f\|_{\infty,D} = \sup_{x \in D} |f(x)|$$

denotes the least upper bound of its absolute value.

Mean value theorems. For a continuously differentiable function f,

$$f(y) - f(x) = f'(z)(y - x)$$

with $z \in [x, y]$.

If f and g are continuous on the closure of a bounded domain D and g is nonnegative, then

$$\int_D f g = f(z) \int_D g$$

for some $z \in D$.

Multi-indices. A multi-index $\alpha = (\alpha_1, \ldots, \alpha_d)$ is a vector of nonnegative integers. It is used to specify partial derivatives and powers of monomials:

$$\partial^\alpha f(x) = \left(\frac{\partial}{\partial x_1}\right)^{\alpha_1} \cdots \left(\frac{\partial}{\partial x_d}\right)^{\alpha_d} f(x), \quad x^\alpha = x_1^{\alpha_1} \cdots x_d^{\alpha_d}.$$

In this context, the notations

$$|\alpha| = \sum_k \alpha_k, \quad \alpha! = \prod_k \alpha_k!$$

are used. Moreover, $\alpha \le \beta$ if $\alpha_k \le \beta_k$ for all k.

Neumann series. If the norm of a square matrix A is less than 1, then $E - A$, with E the unit matrix, is invertible and

$$(E - A)^{-1} = \sum_{k=0}^\infty A^k.$$

Moreover, $\|(E-A)^{-1}\| \le (1-\|A\|)^{-1}$. These formulas apply more generally to linear operators $L: V \to V$ on a vector space V.

For a diagonally dominant matrix D with

$$\sum_{\ell \neq k} |d_{k,\ell}| \le (1-\varepsilon)|d_{k,k}|,$$

application of the Neumann criterion to $E - A = \operatorname{diag}(d_{1,1}, d_{2,2}, \ldots)^{-1} D$ implies that

$$\|D^{-1}\|_\infty \le \frac{1}{\varepsilon} \frac{1}{\min_k d_{k,k}}.$$

Norm. A norm $\|\ \|$ is a real-valued, nonnegative function on a vector space V which satisfies

- $\|u\| > 0$ for $u \neq 0$,

- $\|\lambda u\| = |\lambda| \|u\|$,

- $\|u + v\| \le \|u\| + \|v\|$

for arbitrary vectors u, v and scalars λ.

The standard norm on $\mathbb{R}^d$ is the Euclidean norm, which we denote by single bars:

$$|x| = \left(\sum_{v=1}^{d} |x_v|^2 \right)^{1/2}$$

for $x \in \mathbb{R}^d$.

Parametrization. A parametrization of a curve is a map of an interval to $\mathbb{R}^d$,

$$t \mapsto (p_1(t), \ldots, p_d(t)).$$

It is regular if $p'(t) \neq 0$ for any t. Similarly,

$$(t_1, t_2) \mapsto (p_1(t), p_2(t), p_3(t))$$

parametrizes a surface. Regularity means that the normal vector $\partial_1 p(t) \times \partial_2 p(t)$ never vanishes.

Regularity implies in particular that smoothness of the parametrization is necessary and sufficient for smoothness of the curve or surface, respectively.

We say that a surface is orientable if the normal vector varies smoothly. This is not implied by smoothness of the parametrization. It is a condition which restricts the topological form of the surface.

Points, vectors, and matrices. Points p and vectors v in $\mathbb{R}^d$ are represented as row and column d-tupels of their coordinates, respectively:

$$p = \left(\begin{matrix} p_1, & \ldots, & p_d \end{matrix} \right), \quad v = \left(\begin{matrix} v_1 \\ \vdots \\ v_d \end{matrix} \right).$$

Via transposition we can change between row and column format, i.e.,

$$\left(\begin{array}{ccc} x_1, & \ldots, & x_m \end{array} \right)^{\mathrm{t}} = \left(\begin{array}{c} x_1 \\ \vdots \\ x_m \end{array} \right).$$

An $m \times n$ matrix

$$A = \left(\begin{array}{ccc} a_{1,1} & \cdots & a_{1,n} \\ \vdots & & \vdots \\ a_{m,1} & \cdots & a_{m,n} \end{array} \right)$$

represents a linear map via

$$x \mapsto y = Ax, \quad y_k = \sum_{\ell} a_{k,\ell} x_{\ell}.$$

Projector. A projector $P : V \to U \subseteq V$ is a linear map on a vector space V which leaves the elements of a subspace U invariant, i.e.,

$$Pu = u \; \forall \, u \in U \quad \Leftrightarrow \quad P^2 = P.$$

If V is equipped with a scalar product, orthogonal projections are particularly important. If $\{u_1, \ldots, u_m\}$ is an orthogonal basis for U, then

$$Pv = \sum_{k=1}^{m} \frac{\langle v, u_k \rangle}{\langle u_k, u_k \rangle} u_k$$

is the best approximation from U to v with respect to the induced norm. As a consequence,

$$\langle v - Pv, u \rangle = 0 \quad \forall u \in U.$$

Scalar product. A scalar product,

$$u, v \mapsto \langle u, v \rangle \in \mathbb{R},$$

on a (real) vector space V is a positiv definite, symmetric bilinear map. This means that

- $\langle u, u \rangle > 0$ for $u \neq 0$,

- $\langle u, v \rangle = \langle v, u \rangle$,

- $\langle u, \alpha v + \beta w \rangle = \alpha \langle u, v \rangle + \beta \langle u, w \rangle$

for arbitrary $u, v, w \in V$ and $\alpha, \beta \in \mathbb{R}$. With any scalar product we can associate the norm $\|v\| = \sqrt{\langle v, v \rangle}$.

Square integrable differentiable functions. The space $L_2(D)$ consists of all measurable functions f for which $|f|^2$ is Lebesgue integrable. It is a Hilbert space with the scalar product

$$\langle f, g \rangle = \int_D f g.$$

The functions for which also the first order derivatives are square integrable form the Sobolev space $H^1(D)$ with the scalar product

$$\langle f, g \rangle = \int_D (\operatorname{grad} f)^{\mathrm{t}} \operatorname{grad} g + f g$$

and the associated norm. The subspace of functions which vanish on $\Gamma \subseteq \partial D$ is denoted by $H^1_\Gamma(D)$.

Bibliography

[1] J.H. Ahlberg, E.N. Nielson, and J.L. Walsh: *The Theory of Splines and Their Applications*, Academic Press, 1967. (Cited on p. ix)

[2] A.G. Aitken: *On interpolation by iteration of proportional parts, without the use of differences*, Proc. Edinburgh Math. Soc. 3 (1932), 56–76. (Cited on p. 193)

[3] R.E. Barnhill and R.F. Riesenfeld (eds.): *Computer-Aided Geometric Design*, Academic Press, 1974. (Cited on p. ix)

[4] B.A. Barsky: *The β-spline, a local representation based on shape parameters and fundamental geometric measures*, PhD thesis, Dept. of Computer Science, University of Utah, 1981. (Cited on p. 193)

[5] B.A. Barsky and T.D. DeRose: *An intuitive approach to geometric continuity for parametric curves and surfaces*, in Computer Generated Images: The State of the Art, N. Magnenay and D. Thalmann (eds.), Springer (1985), 159–175. (Cited on p. 193)

[6] S. Bernstein: *Démonstration du théorème de Weierstrass, fondée sur le calcul des probabilités*, Comm. Soc. Math. Kharkow 13 (1912–1913), 1–2. (Cited on p. 193)

[7] P. Bézier: *Définition numérique des courbes et surfaces I*, Automatisme XI (1966), 625–632. (Cited on pp. ix, 193)

[8] P. Bézier: *Définition numérique des courbes et surfaces II*, Automatisme XII (1967), 17–21. (Cited on pp. ix, 193)

[9] P. Bézier: *Procédé de définition numérique des courbes et surfaces non mathématiques*, Automatisme XIII/5 (1968), 189–196. (Cited on pp. ix, 193)

[10] P. Bézier: *Numerical Control: Mathematics and Applications*, John Wiley & Sons, 1972. (Cited on pp. ix, 193)

[11] W. Boehm: *Inserting new knots into B-spline curves*, Computer-Aided Design 12 (1980), 199–201. (Cited on pp. ix, 194)

[12] W. Boehm: *Curvature continuous curves and surfaces*, Computer Aided Geometric Design 2 (1985), 313–323. (Cited on p. 193)

[13] W. Boehm and H. Prautzsch: *Geometric Concepts for Geometric Design*, AK Peters, 1994. (Cited on pp. 193, 197)

[14] C. de Boor: *On uniform approximation by splines*, J. Approx. Theory 1 (1968), 219–235. (Cited on p. 194)

[15] C. de Boor: *On calculating with B-splines*, J. Approx. Theory 6 (1972), 50–62. (Cited on pp. ix, 194)

[16] C. de Boor: *Total positivity of the spline collocation matrix*, Ind. Univ. J. Math. 25 (1976), 541–551. (Cited on p. 194)

[17] C. de Boor: *Splines as linear combinations of B-splines*, in Approximation Theory II, G.G. Lorentz, C.K. Chui, and L.L. Schumaker (eds.), Academic Press, New York, 1976, 1–47. (Cited on pp. ix, 194)

[18] C. de Boor: *Package for calculating with B-splines*, SIAM J. Numer. Anal. 14 (1977), 441–472. (Cited on pp. ix, 194)

[19] C. de Boor: *A Practical Guide to Splines*, Springer, 1978. (Cited on pp. ix, 194)

[20] C. de Boor and G.J. Fix: *Spline approximation by quasi-interpolants*, J. Approx. Theory 8 (1973), 19–45. (Cited on p. 194)

[21] C. de Boor and K. Höllig: *B-splines without divided differences*, in: Geometric Modeling: Algorithms and New Trends, G.E. Farin (ed.), SIAM (1987), 21–27. (Cited on p. 194)

[22] C. de Boor, K. Höllig, and S. Riemenschneider: *Box Splines*, Springer, 1993. (Cited on pp. ix, 194)

[23] C. de Boor, K. Höllig, and M. Sabin, *High accuracy geometric Hermite interpolation*, Computer Aided Geometric Design 4 (1987), 269–278. (Cited on p. 193)

[24] C. de Boor and L.L. Schumaker: *Spline Bibliography* http://pages.cs.wisc.edu/~deboor/bib/bib.html, website, 2011. (Cited on p. xi)

[25] P. de Casteljau: *Outillages méthodes calcul*, Technical report, André Citroën Automobiles SA, Paris, 1959. (Cited on pp. ix, 193)

[26] P. de Casteljau: *Courbes et surfaces à pôles*, Technical report, André Citroën Automobiles SA, Paris, 1963. (Cited on pp. ix, 193)

[27] E. Catmull and J. Clark: *Recursively generated B-spline surfaces on arbitrary topological meshes*, Computer Aided Design 10/6 (1978), 350–355. (Cited on pp. ix, 195)

[28] J. Céa: *Approximation Variationelle des Problèmes aux limites*, PhD thesis, Annales de l'institut Fourier 14.2. (1964), 345–444. (Cited on p. 195)

[29] G.M. Chaikin: *An algorithm for high-speed curve generation*, Computer Graphics and Image Processing 3 (1974), 346–349. (Cited on pp. ix, 194)

[30] E. Cohen, T. Lyche, and R.F. Riesenfeld: *Discrete B-splines and subdivision techniques in computer-aided geometric design and computer graphics*, Computer Graphics Image Proc. 14 (1980), 87–111. (Cited on pp. ix, 194)

[31] E. Cohen, R.F. Riesenfeld, and G. Elber: *Geometric Modeling with Splines: An Introduction*, AK Peters, 2001. (Cited on pp. ix, 194, 195)

[32] E. Conte and C. de Boor: *Elementary Numerical Analysis*, McGraw-Hill, 1972. (Cited on pp. 193, 197)

[33] S. Coons: *Surfaces for Computer-Aided Design*, Technical Report, M.I.T., 1964. (Cited on pp. ix, 193, 195)

[34] J.A. Cottrell, T.J.R. Hughes, and Y. Bazilevs: *Isogeometric Analysis: Toward Integration of CAD and FEA*, Wiley, 2009. (Cited on pp. x, 195)

Bibliography

[35] M.G. Cox: *The numerical evaluation of B-splines*, J. Inst. Math. Appl. 10 (1972), 134–149. (Cited on p. 194)

[36] W. Dahmen and C.A. Micchelli: *Some results on box splines*, Bull. Amer. Math. Soc. 11 (1984), 147–150. (Cited on p. ix)

[37] I. Daubechies, I. Guskov, and W. Sweldens: *Regularity of irregular subdivision*, Constr. Approx. 15 (1999), 381–426. (Cited on p. 193)

[38] O. Davydov, J. Prasiswa, and U. Reif: *Two-stage approximation methods with extended B-splines*, Mathematics of Computation (to appear). (Cited on p. 195)

[39] T.D. DeRose: *Geometric continuity: a parametrization independent measure of continuity for computer aided geometric design*, PhD thesis, University of California, Berkeley, 1985. (Cited on p. 193)

[40] R.A. DeVore, W. Dahmen, and K. Scherer: *Multidimensional spline approximation*, SIAM J. Numer. Anal. 17 (1980), 380–402. (Cited on pp. 194, 195)

[41] R.A. DeVore and G.G. Lorentz: *Constructive Approximation*, Springer, 1993. (Cited on p. 193)

[42] T. Dokken, T. Lyche, and K.F. Petterson: *Polynomial splines over locally refined box-partitions*, Computer Aided Geometric Design 30 (2013), 331–356. (Cited on p. 195)

[43] D. Doo and M. Sabin: *Behaviour of recursive subdivision surfaces near extraordinary points*, Computer-Aided Design 10/6 (1978), 356–360. (Cited on pp. ix, 195)

[44] S. Dubuc: *Interpolation through an iterative scheme*, J. Math. Anal. Appl. 114 (1986), 185–204. (Cited on p. 193)

[45] N. Dyn, D. Levin, and J. Gregory: *A 4-point interpolatory subdivision scheme for curve design*, Computer Aided Geometric Design 4 (1987), 257–268. (Cited on p. 193)

[46] I.D. Faux and M.J. Pratt: *Computational Geometry for Design and Manufacture*, Wiley Ellis Horwood Ltd., 1979. (Cited on p. ix)

[47] G.E. Farin: *Algorithms for rational Bézier curves*, Computer Aided Geometric Design 15 (1983), 73–77. (Cited on pp. 193, 194)

[48] G.E. Farin (ed.): *Geometric Modeling: Algorithms and New Trends*, SIAM, 1987. (Cited on p. ix)

[49] G.E. Farin: *Curves and Surfaces for Computer Aided Geometric Design*, Academic Press, New York, 1988, fifth edition, 2002. (Cited on pp. ix, 193, 194)

[50] G.E. Farin (ed.): *NURBS for Curve and Surface Design*, SIAM, 1991. (Cited on p. 193)

[51] G.E. Farin: *NURB Curves and Surfaces*, AK Peters, 1999. (Cited on p. 193)

[52] G. Farin, J. Hoschek, and M.-S. Kim (eds.): *Handbook of Computer Aided Geometric Design*, North Holland, 2002. (Cited on p. ix)

[53] J. Ferguson: *Multivariable curve interpolation*, JACM, 11/2 (1964), 221–228. (Cited on pp. ix, 194)

[54] M.S. Floater: *A piecewise polynomial approach to analyzing interpolatory subdivision*, J. Approx. Theory 163 (2011), 1547–1563. (Cited on p. 193)

[55] A.R. Forrest: *Curves and Surfaces for Computer-Aided Design*, PhD thesis, Cambridge University, 1968. (Cited on p. 193)

[56] D.R. Forsey and R.H. Bartels: *Hierarchical B-spline refinement*, Computer Graphics 22 (1988), 205–212. (Cited on p. 195)

[57] B.G. Galerkin: *Stäbe und Platten; Reihen in gewissen Gleichgewichtsproblemen elastischer Stäbe und Platten*, Vestnik der Ingenieure 19 (1915), 897–908. (Cited on p. 195)

[58] W.J. Gordon: *Blending function methods of bivariate and multivariate interpolation and approximation*, SIAM J. Numer. Anal. 8 (1971), 158–177. (Cited on p. 195)

[59] W.J. Gordon and R.F. Riesenfeld: *B-spline curves and surfaces*, in: Computer Aided Geometric Design, R.E. Barnhill and R.F. Riesenfeld (eds.), Academic Press (1974), 95–126. (Cited on p. 194)

[60] T. A. Grandine and T. A. Hogan: *A parametric quartic spline interpolant to position, tangent, and curvature*, Computing 72 (2004), 65–78. (Cited on p. 193)

[61] I. Grieger: *Graphische Datenverarbeitung*, Springer, 1987. (Cited on p. 195)

[62] C. Hermite: *Sur la formule d'interpolation de Lagrange*, J. Reine Angew. Math. 84 (1878), 70–79. (Cited on p. 193)

[63] J.C. Holladay: *Smoothest curve approximation*, Math. Tables Aids Comput. 11 (1957), 233–243. (Cited on p. 194)

[64] K. Höllig: *Stability of the B-spline basis via knot insertion*, Computer Aided Geometric Design 17 (2000), 447–450. (Cited on p. 194)

[65] K. Höllig: *Finite Element Methods with B-Splines*, SIAM, 2003. (Cited on pp. x, 195)

[66] K. Höllig and J. Hörner: *Finite Element Methods with B-Splines: Supplementary Material*, http://www.siam.org/books/fr26/, website, 2012. (Cited on p. 195)

[67] K. Höllig, J. Hörner, and A. Hoffacker: *Finite element analysis with B-splines: weighted and isogeometric methods*, in: Curves and Surfaces 2011, J.D. Boissonnat et al. (eds.), LNCS 6920, Springer (2012), 330–350. (Cited on p. 195)

[68] K. Höllig and J. Koch: *Geometric Hermite interpolation with maximal order and smoothness*, Computer Aided Geometric Design 13 (1996), 681–695. (Cited on p. 193)

[69] K. Höllig, U. Reif, and J. Wipper: *Weighted extended B-spline approximation of Dirichlet problems*, SIAM J. Numer. Anal. 39 (2001), 442–462. (Cited on pp. x, 195)

[70] J. Hoschek and D. Lasser: *Fundamentals of Computer Aided Geometric Design*, AK Peters, 1993. (Cited on pp. 193, 194, 195)

[71] T.J.R. Hughes, J.A. Cottrell, and Y. Bazilevs: *Isogeometric analysis: CAD, finite elements, NURBS, exact geometry and mesh refinement*, Computer Methods in Applied Mechanics and Engineering 194 (2005), 4135–4195. (Cited on pp. x, 195)

[72] G. Jaklič, J. Kozak, M. Krajnc, V. Vitrih, and E. Žagar: *Hermite geometric interpolation by rational Bézier spatial curves*, SIAM J. Numer. Anal. 50 (2012), 2695–2715. (Cited on p. 193)

[73] L.W. Kantorowitsch and W.I. Krylow: *Näherungsmethoden der Höheren Analysis*, VEB Deutscher Verlag der Wissenschaften, 1956. (Cited on p. 195)

[74] W.A. Kincaid and E.W. Cheney: *Numerical Analysis: Mathematics of Scientific Computing*, American Mathematical Society, 2002. (Cited on pp. 193, 197)

[75] R. Kraft: *Adaptive und linear unabhängige multilevel B-Splines und ihre Anwendungen*, Dissertation, Stuttgart, 1998. (Cited on p. 195)

Bibliography 209

[76] M. Lai and L.L. Schumaker: *Spline Functions on Triangulations*, Cambridge University Press, 2007. (Cited on p. 194)

[77] J.M. Lane and R.F. Riesenfeld: *A theoretical development for the computer generation and display of piecewise polynomial surfaces*, IEEE Trans. Pattern Anal. Mach. Intellig. 2 (1980), 35–45. (Cited on p. 194)

[78] J.M. Lane and R.F. Riesenfeld: *A geometric proof for the variation diminishing property of B-spline approximation*, J. Approx. Theory 37 (1983), 1–4. (Cited on p. 194)

[79] E.T.Y. Lee: *The rational Bézier representation of conics*, in: Geometric Modeling: Algorithms and New Trends, G.E. Farin (ed.), SIAM (1987), 3–19. (Cited on p. 193)

[80] G.G. Lorentz: *Bernstein Polynomials*, Toronto Press, 1953, Second Edition, 1986. (Cited on p. 193)

[81] T. Lyche: *Private communication*, Avignon, 2010. (Cited on p. 195)

[82] T. Lyche and L.L. Schumaker: *Local spline approximation methods*, J. Approx. Theory 15 (1975), 294–325. (Cited on p. 194)

[83] J. Marsden: *An identity for spline functions with applications to variation-diminishing spline approximation*, J. Approx. Theory 3 (1970), 7–49. (Cited on p. 194)

[84] The MathWorks, Inc.: MATLAB – *The Language of Technical Computing*, http://www.mathworks.com/products/matlab/index.html, website, 2011. (Cited on p. xi)

[85] E.H. Neville: *Iterative interpolation*, J. Indian Math. Soc. 20 (1934), 87–120. (Cited on p. 193)

[86] J. Peters and U. Reif: *Subdivision Surfaces*, Springer, 2008. (Cited on pp. x, 195)

[87] L. Piegl and W. Tiller: *The NURBS Book*, Springer, 1997. (Cited on pp. 194, 195)

[88] A. Pinkus: *n-Widths in Approximation Theory*, Springer, 1985. (Cited on p. ix)

[89] H. Prautzsch: *Free-form splines*, Computer Aided Geometric Design 14 (1997), 201–206. (Cited on p. 195)

[90] H. Prautzsch, W. Boehm, and M. Paluszny: *Bézier and B-Spline Techniques*, Springer, 2002. (Cited on pp. ix, 195)

[91] A. Rababah: *High order approximation method for curves*, CAGD 12 (1995), 89–102. (Cited on p. 193)

[92] L. Ramshaw: *Blossoming: A Connect-the-Dots Approach to Splines*, Technical Report, Digital Systems Research Center, Paolo Alto, 1987. (Cited on p. 194)

[93] U. Reif: *Neue Aspekte in der Theorie der Freiformflächen beliebiger Topologie*, Dissertation, Universität Stuttgart, 1993. (Cited on p. 195)

[94] U. Reif: *Orthogonality of B-splines in weighted Sobolev spaces*, SIAM J. Math. Anal. 28 (1997), 1258–1263. (Cited on p. 194)

[95] U. Reif: *TURBS: Topologically unrestricted rational B-splines*, Constructive Approximation 14 (1998), 57–78. (Cited on p. 195)

[96] U. Reif: *Best bounds on the approximation of polynomials and splines by their control structure*, Computer Aided Geometric Design 17 (2000), 579–589. (Cited on p. 194)

[97] U. Reif: *Polynomial approximation on domains bounded by diffeomorphic images of graphs*, Journal of Approximation Theory 164 (2012), 954–970. (Cited on p. 194)

[98] H. Reinsch: *Smoothing by spline functions*, Numer. Math. 10 (1967), 177–183. (Cited on p. 194)

[99] R.F. Riesenfeld: *Applications of B-Spline Approximation to Geometric Problems of Computer-Aided Design*, PhD thesis, Syracuse University, 1973. (Cited on p. 194)

[100] R.F. Riesenfeld: *On Chaikin's algorithm*, IEEE Computer Graphics and Applications 4 (1975), 304–310. (Cited on pp. ix, 194)

[101] W. Ritz: *Über eine neue Methode zur Lösung gewisser Variationsprobleme der mathematischen Physik*, J. Reine Angew. Math. 135 (1908), 1–61. (Cited on p. 195)

[102] C. Runge: *Über empirische Funktionen und die Interpolation zwischen äquidistanten Ordinaten*, Zeitschrift für Mathematik und Physik 46 (1901), 224–243. (Cited on p. 193)

[103] V.L. Rvachev and T.I. Sheiko: *R-functions in boundary value problems in mechanics*, Appl. Mech. Rev. 48 (1995), 151–188. (Cited on p. 195)

[104] R. Schaback: *Optimal geometric Hermite interpolation of curves*, in: Mathematical Methods for Curves and Surfaces II (1998), M. Daehlen, T. Lyche, and L. Schumaker (eds.), Vanderbilt University Press (1998), 1–12. (Cited on p. 193)

[105] I.J. Schoenberg: *Contributions to the problem of approximation of equidistant data by analytic functions*, Quart. Appl. Math. 4 (1946), 45–99 and 112–141. (Cited on pp. ix, 194)

[106] I.J. Schoenberg: *Spline functions and the problem of graduation*, Proc. Nat. Acad. Sci. 52 (1964), 947–950. (Cited on p. 194)

[107] I.J. Schoenberg: *Cardinal Spline Interpolation*, CMBS, SIAM, 1973. (Cited on p. 194)

[108] I.J. Schoenberg and T.N.E. Greville: *On spline functions*, in: Inequalities, O. Shisha (ed.), Academic Press (1967), 255–291. (Cited on p. 194)

[109] I.J. Schoenberg and A. Whitney: *On Pólya frequency functions III: The positivity of translation determinants with application to the interpolation problem by spline curves*, Trans. Amer. Math. Soc. 74 (1953), 246–259. (Cited on p. 194)

[110] M.H. Schultz: *Rayleigh–Ritz–Galerkin methods for multidimensional problems*, SIAM J. Numer. Anal. 4 (1969), 523–538. (Cited on p. 195)

[111] L.L. Schumaker: *Spline Functions: Basic Theory*, Wiley-Interscience, 1980. (Cited on pp. ix, 194, 195)

[112] T.W. Sederberg, J. Zheng, A. Bakenov, and A. Nasri: *T-splines and T-NURCCS*, ACM Transactions on Graphics 22 (2003), 477–484. (Cited on p. 195)

[113] E. Staerk: *Mehrfach differenzierbare Bézierkurven und Bézierflächen*, PhD thesis, TU Braunschweig, 1976. (Cited on p. 193)

[114] G. Strang: *Introduction to Linear Algebra*, Wellesley-Cambridge Press, 1993, Fourth Edition 2009. (Cited on p. 197)

[115] G. Strang: *Linear Algebra and Its Applications*, Cengage, 1976, Fourth Edition 2006. (Cited on p. 197)

[116] G. Strang: *Computational Science and Engineering*, Wellesley-Cambridge Press, 2007. (Cited on p. 193)

Bibliography

[117] G. Strang and G.J. Fix: *An Analysis of the Finite Element Method*, Prentice–Hall, Englewood Cliffs, NJ, 1973, Second Edition, Wellesley-Cambridge Press, 2008. (Cited on pp. 194, 195, 197)

[118] B. Taylor: *Methodus Incrementorum Directa et Inversa*, London: Gul. Innys, 1715. (Cited on p. 193)

[119] W. Tiller: *Rational B-splines for curve and surface representations*, IEEE Computer Graphics & Applications 3 (1983), 61–69. (Cited on p. 194)

[120] K.J. Verspille: *Computer-Aided Design Applications of the Rational B-Spline Approximation Form*, PhD Thesis, Syracuse University, 1975. (Cited on p. 194)

[121] E. Waring: *Problems concerning interpolations*, Philos. Trans. R. Soc. Lond. 69 (1779), 59–67. (Cited on p. 193)

[122] K. Weierstrass: *Über die analytische Darstellbarkeit sogenannter willkürlicher Funktionen einer reellen Veränderlichen*, Sitzungsberichte der Königlich Preußischen Akademie der Wissenschaften zu Berlin, 1885, I 633–639 and II 789–805. (Cited on p. 193)

Index

affine invariance, 38
AITKEN–NEVILLE scheme, 7
approximation, tensor product, 146

B-spline, 52
    multivariate, 138
    stability, 88
    uniform, **66**, 139
    weighted, 177
basis, 197
BAZILEVS, 180
BERNSTEIN polynomial, 9
BÉZIER
    curve, 19
    patch, 155
    rational curve, 37
    surface, 158
bilinear form, 197
blossoming, 129
BOEHM, 115
DE BOOR, 34, 52, 82, 88, 91, 140
    algorithm, 68
bounding box, 23, **124**
box-spline, 140

CATMULL–CLARK algorithm, 165
CAUCHY–SCHWARZ inequality, 197
CEA lemma, 176
COHEN, 115
conic section, 46, **197**
control polygon, 19
convex combination, 198
COONS, 37
    patch, 166
COTRELL, 180
COX, 52
cross product, 198
curvature, 29, **198**

DE CASTELJAU algorithm, 25
DEVORE, 140
diagonally dominant, 199
differentiation
    B-spline, 55
    BÉZIER curve, 27
    multivariate spline, 144
    rational BÉZIER curve, 44
    spline, 71
    spline curve, 125
    uniform spline, 73
distance, 112
divided difference, 199
DOKKEN, 151
DUBUC, 7

endpoint interpolation, **22**, 110
error
    interpolation, 95
    polynomial approximation, 137
    spline approximation, 148
evaluation
    multivariate spline, 142
    spline, 68
    spline curve, 125

FARIN, 38
finite element, 173
FIX, 82, 197
FORREST, 37

GORDON, 105

hierarchical splines, 151
HILBERT space, 199
HOLLADAY, 97
homogeneous coordinates, 42, **199**
HOSCHECK, 168
HUGHES, 180

incompressible flow, 187
interpolation
    geometric, 34
    HERMITE, 14
    natural, 97
    spline curve, 130
isogeometric element, 180

KANTOROWITSCH, 177
knot insertion, 116
knot sequence, 51
KRONECKER symbol, 200
KRYLOW, 177

LAME–NAVIER system, 190
LANDAU symbol, 200
LANE, 123
LASSER, 168
least squares line, 200
linear elasticity, 190
linear functional, 201
linear interpolation, 200
LYCHE, 82, 115, 151

MANSFIELD, 52
MARSDEN identity, 59
maximum norm, 201
mean value theorem, 201
multiindex, 201

nested multiplication, 3
NEUMANN series, 201
norm, 202

parametrization, 202
PETERS, 166
PETTERSON, 151
polynomial
    BERNSTEIN, **9**, 53
    LAGRANGE form, 6
    monomial form, 1

multivariate, 133
  BERNSTEIN, 135
  TAYLOR, 3
projector, 203
  standard, 84

quasi-interpolant, 82
  accuracy, 86

R-function, 178
RAMSHAW, 129
recurrence relation, 52
REIF, 81, 111, 137, 160, 166, 180
REINSCH, 100
RIESENFELD, 105, 115, 123
RITZ–GALERKIN
  approximation, 174
  system, 184

RUNGE, 8
RVACHEV, 178

SABIN, 34
scalar product, 203
SCHOENBERG, 66, 100
  scheme, 77
SCHOENBERG–WHITNEY
    conditions, 91
SCHULTZ, 177
SCHUMAKER, 82
singular parametrization, 160
smoothing spline, 101
spline, 63
  BÉZIER form, 127
  multivariate, 140
  periodic, 73

rational curve, 108
  solid, 169
spline curve, 105
spline surface, 160
square integrable, 203
STRANG, 197
subdivision
  BÉZIER curve, 31
  uniform, 119

variation diminution, 122

WEIERSTRASS approximation
    theorem, 16
weight points, 38
WIPPER, 180